MR. WIZARD'S EXPERIMENTS
FOR YOUNG SCIENTISTS

Mr. Wizard's EXPERIMENTS for Young Scientists

by Don Herbert

Illustrated by Dan Noonan

DOUBLEDAY & COMPANY, INC.
GARDEN CITY, NEW YORK

ISBN: 0-385-07798-X TRADE
0-385-04540-9 PREBOUND

Library of Congress Catalog Card Number 59-7907

Printed in the United States of America

15 14 13 12 11 10

ACKNOWLEDGMENTS

A book like this is never the work of one person. While I prepared the material and wrote it, I am indebted to hundreds of scientists, teachers, and writers who have supplied what has become my "background," not only in science, but in the collection of facts and attitudes toward them that are a part of my general knowledge. To these extremely important but unknown men and women I owe more than I can ever say.

Another group of people helped me directly as I put the book together. For their generous co-operation I am especially grateful. This group is headed by Dr. Morris Shamos, professor of physics, Washington Square campus, New York University. He helped in many ways: reading manuscripts, suggesting ideas, pointing out short cuts and improvements, and submitting chapters to his colleagues at New York University.

Other scientists suggested improvements or commented on the chapters that covered their fields: Dr. R. Bobrow of the Atlantic Yeast Company and Walter Roider of the Borden Company.

Amateur scientists also contributed valuable ideas. Among them were Ann Hesse and Richard Hohol.

Special mention, too, must be made of the contributions made by Lois Brandon and Frank Zelinski who have worked hard for so many years behind the scenes, not only of the book, but of the "Watch Mr. Wizard" television program.

This book is dedicated to

JEFF, JAY, JILL,

and all the other possible future scientists
to whom the world of tomorrow belongs.

Contents

Introduction

I hope you'll find the experiments in this book fun to do. I've certainly enjoyed the many hours I've spent working with science. These experiments are the best I could devise to help you share that enjoyment.

Each chapter will introduce you to a different scientist. The experiments within the chapters will give you an idea of what that particular kind of scientist does. By the time you've finished all of the experiments, you will have a general idea of what it means to be a scientist and how he goes about his work.

Some of the experiments will probably appeal to you more than others. I hope you'll try them all, because it's only after you have tried the experiment that you can really make up your mind how you feel about that side of science.

To duck something because it seems "hard" is a natural reaction, especially if you're supposed to be enjoying yourself. But remember, you don't ever really get something for nothing! The more you put into these experiments, the more you'll get out of them.

If some of the experiments seem involved, read the directions slowly and carefully. Repeat parts of it until you understand it before you go on to the next part. Things that look complicated often become amazingly simple if you move ahead only one step at a time.

Science is the systematic investigation and explanation of the world around us. In order to organize their work and its results, scientists have divided science into major fields which you call physics, chemistry, and biology. Physics is the study of matter and energy and how they affect one another. Chemistry is the study of the kinds of matter and how they change and are related. Biology is the study of living things. These are the branches of science most often talked about.

Even though it's convenient to think of physics, chemistry, and biology as separate studies, the world itself knows of no such boundaries. Animals, plants, minerals, air, planets, and the hundreds of thousands of other things in the universe are a mixture of physics, chemistry, and biology that cannot really be separated at all.

A fish lives in this complicated interrelated world. This fish may be studied by a biologist to find out what it eats; by a chemist to find out what it is made of; or by a physicist to find out how it swims. The biologist soon realizes he's got to know a lot about physics and chemistry in order to understand what a fish eats and why. The chemist has to know about the biology of the fish and the physics of, say, molecules and atoms to discover what the fish is made of. The physicist must know chemistry and biology to understand completely the mechanisms that make the fish swim.

This interweaving of one branch of science with another is encountered in most scientific investigations. Today most scientists need a broad scientific background even though their work of the moment may be very specialized.

Specialization is necessary as scientists get to know more and more about the world. For example, so much information has been collected about fish in general, a biologist may find it necessary to concentrate on one kind of fish to add to the body of knowledge. The physicist may specialize in atomics or solid-state physics, while the chemist may work with only one group of chemicals.

Sometimes "physicist," "chemist," and "biologist" don't give a specific enough indication of what a scientist does. So names like biochemist and physical chemist are used. You can combine almost any two names of scientific fields and you'll have a word that describes a scientist.

In other cases, the work of the scientist includes so many fields he is given a name that doesn't really relate to the basic science at all. Geologist is one such name. A geologist investigates how the earth and the things on it have come to be the way they are. He will look into the physical, chemical, and biological aspects of the part of the earth he specializes in.

All of these scientists, no matter what they are named, proceed in the same way to achieve their results.

As you work in each of the fields of science, you will want to use their methods to make sure you are proceeding properly. Here is the way a scientist generally works:

1. The scientist organizes the problem he is trying to solve so he can see clearly what he is trying to prove or disprove.
2. He then investigates all the possible explanations for the facts he has found or ways of solving his problem.
3. He selects the one that he thinks has the best chance of being correct.
4. He does experiments to test his tentative explanation or solution to see if it is correct.
5. If the results of the experiment seem to indicate that the explanation or solution is correct, he does more experiments to test it further.
6. He then formulates his explanation or solution so that it will describe what he has found.
7. He explains his work to other scientists so they can verify his results.

The scientist has developed methods of working that help him reduce the number of errors that can be made. Here are some of those methods to help you as you do the experiments in this book.

First, read the whole chapter to be sure you understand what the experiment is about and how you are to proceed. It's important that you know what you're trying to do before you start out, because then you'll be prepared for each step as it comes.

When you gather the material for the experiment, try to use whatever materials are easily available. If quart bottles are called for in the experiment and you have only pint bottles available, examine the conditions of the experiment to see if the smaller size of the bottle will substantially change the conditions. In some cases, following the instructions is very important; in others, the instructions need only be used as a guide.

Most of the equipment can be found around the house or purchased at a drug, hardware, grocery, or variety store. Most of the chemicals called for are available in the pantry or medicine cabinet. Those that aren't can be purchased at a hardware store, or drugstore.

Do the steps of the experiments carefully to be sure each part has been done properly and nothing has been omitted. One way to do this is to make a check list beforehand and then do a "count down" as you finish each step. This check list can also be used as an outline for the records which you should keep. Keeping accurate and complete records is important, because you may forget what you did after a week has gone by. You may also overlook such facts as an increase in temperature since your last observation that may affect your results.

In general, it's a good idea to include information like the following if it can at all affect the results of the experiment: date, time, place, temperature, pressure, size, shape, kinds, color, amount, weight, length, width, height, number, and order.

Try to keep the condition under which the experiment is done as constant as possible. Then vary only one thing at a time. For example, when you enter the field of zoology and run your zoo, keep your animals in the same place, at the same temperature, give them the same amount of water, and in every other way try to keep the conditions of their environment constant. Then feed them only one food at a time . . . at the same time every day. A serious zoologist might weigh the food before and after each day to determine exactly how much food the animals ate. If there is any change in their behavior during this period, one of the first places to look for an explanation of the new behavior would be the food.

When you have finished the experiments, perhaps you will want to do other experiments in the same field. You can probably think of others to try using the same equipment.

You don't have to have complicated equipment to do interesting experiments. If you've read about Mendel and the studies he made with garden peas, you know that he needed only several kinds of garden peas, the right conditions in which to grow them, and accurate records to discover a scientific law that helps us understand how we inherit characteristics, such as blue eyes or brown eyes, from our parents.

Be careful that you don't jump to unwarranted conclusions. When you've done the experiment and have recorded your results, do it again to see if you get the same results. Repeat this process several times to see if the results continue to be the same. If there is a difference, try to account for it. Sometimes the differences occur because the experimenter is not careful in his procedure. The more exact you are, the more you can rely on your results.

Ask yourself, "Why is this happening?" at every step of the

experiment. You'll be surprised at how much you know about some things and how little even experienced scientists know about others. Even though scientists have found out a lot about the world, there is still much to be discovered.

After you've finished doing all the experiments in this book, you may want to continue experimenting in those branches of science that especially interested you. Be sure to look up your "specialities" in the library. If you found that you enjoyed growing crystals, for instance, you will find there are whole books on nothing but crystals. Eventually you will want to have a library of science books of your own. Lifetime hobbies have developed from experiments like those in the following pages. Who can tell—you may find one yourself.

It's even possible you will discover that you enjoy one of the chapters so much you will decide to make the subject it covers your lifework. If so, you have a rich life to look forward to, because scientists are the busiest, happiest, most stimulating, and most successful people I know.

MR. WIZARD'S EXPERIMENTS FOR YOUNG SCIENTISTS

1. Measuring the Sun

Astronomer

Do you know what the diameter of the sun is? Why not figure it out for yourself without even a telescope? At the same time you'll see how an astronomer finds out about the planets, sun, and stars.

First, draw two parallel lines one inch apart on a piece of white cardboard and fold as shown:

Select a room that has a window facing the sun at the time of day when you are going to do the experiment. Pull down the shades in the room to get it as dark as you can. On the window facing the sun, pull down the shade only half way. Cut a section out of the center of a piece of corrugated cardboard and mount a thinner piece of cardboard over this section. Make a hole with a pin through the center of the thinner cardboard and place the whole thing in the lower half of the window facing the sun. When you get it into place it will look like this:

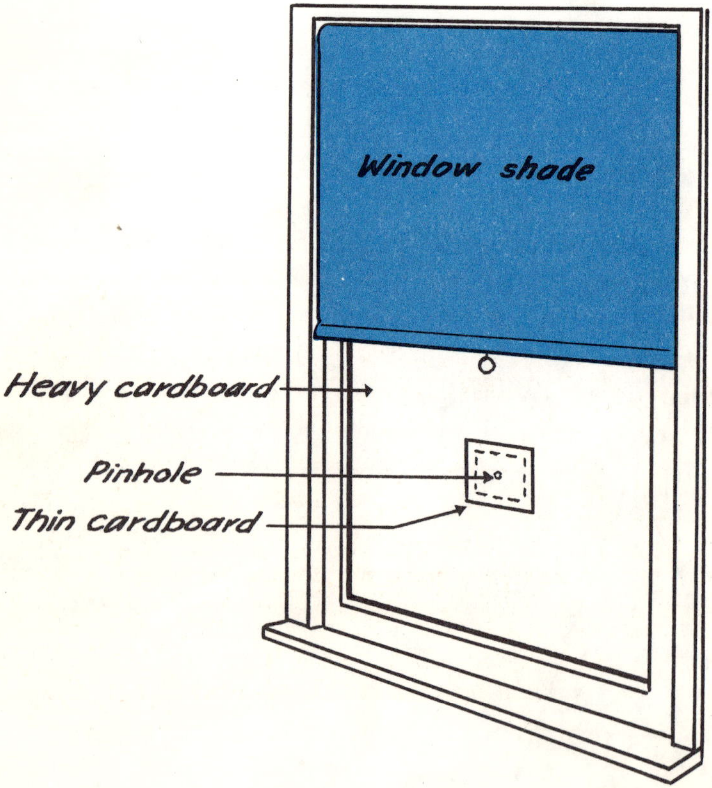

Set up the cardboard that you have marked with the lines an inch apart so the sunlight coming through the pinhole falls directly on it. You will see a small image of the sun! Adjust the front side of the cardboard to get it at right angles to the beam of light. Move the cardboard until the image of the sun falls just within the two parallel lines. You will now have an image of the sun one inch in diameter.

Measure the distance from the image of the sun on the cardboard to the pinhole at the window as accurately as you can.

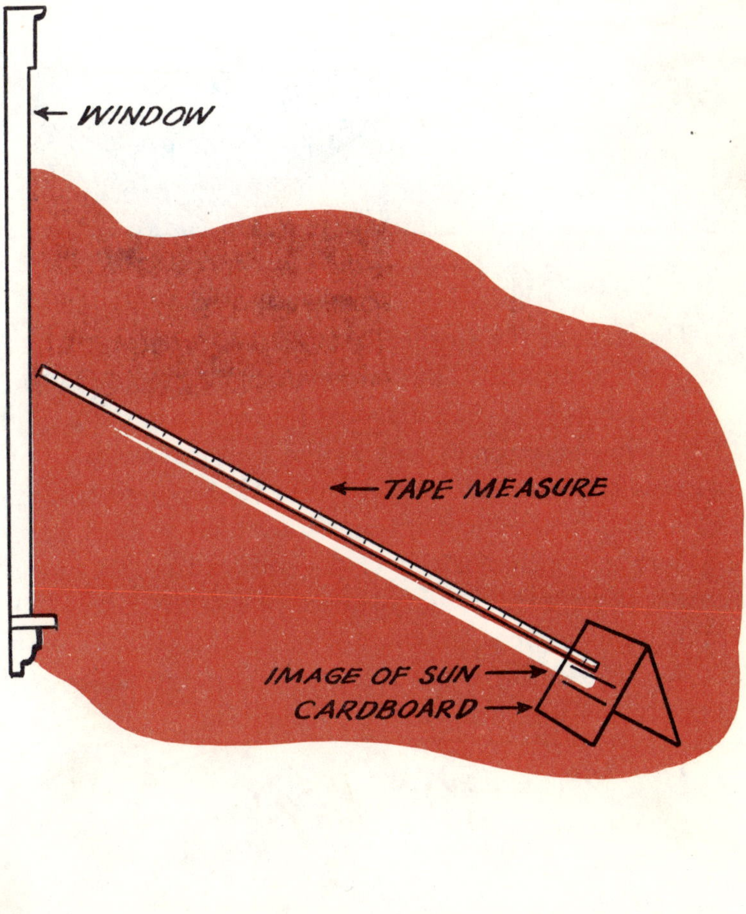

With these measurements you are ready to figure out the diameter of the sun. To do this you must first understand how mathematicians can find an unknown distance by comparing imaginary triangles. This process is an easy one to follow. Here you can see the two triangles are the same shape but not the same size:

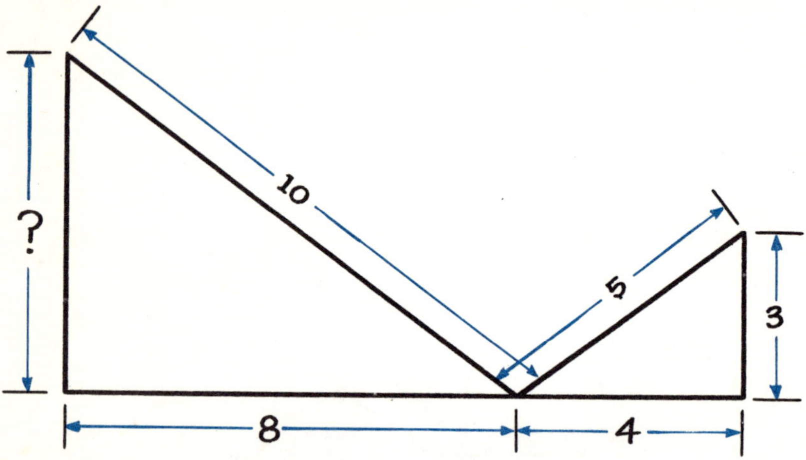

The base of the smaller triangle is four inches, while the corresponding line in the larger triangle is eight inches . . . just twice as long. The diagonal side of the larger triangle is ten inches or twice as long as the five-inch diagonal of the smaller triangle. The vertical side of the larger triangle is unknown, but mathematicians know that it must be twice as long as the similar side of the smaller triangle. Since this side is three inches, the unknown side of the larger triangle must be six inches.

This relationship of the sides to one another can be written in mathematical shorthand or a formula as:

$$4/8 = 5/10 = 3/? \qquad ? = 6$$

Note that the numbers that correspond to the sides of the smaller triangle are above the line while the numbers that represent the value of the sides of the larger triangle are below the line.

The same relationships exist even though the similar triangles are in different positions. They could be as shown like this, for example, and the unknown side would still be six inches.

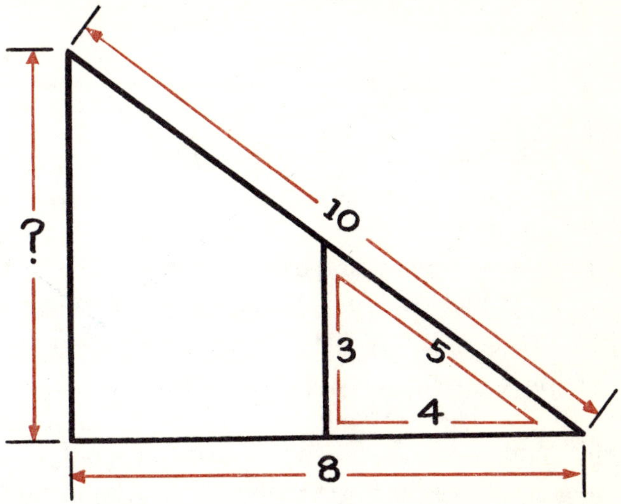

With this knowledge you can now construct two triangles based on the measurements you have made in the dark room.

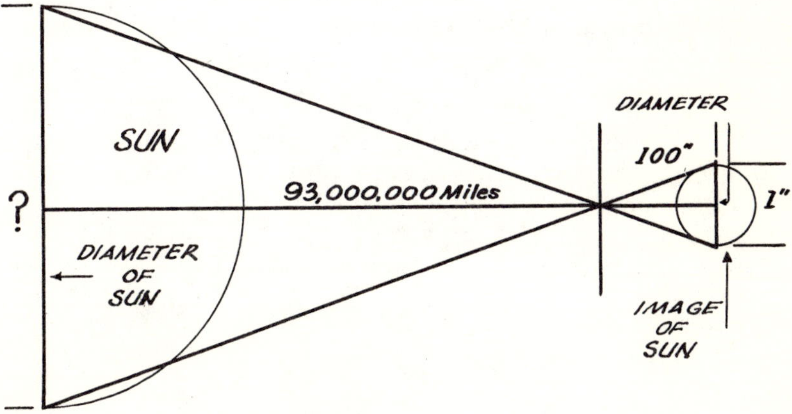

By referring to the figure above, you can see that the imaginary lines form our triangles. The distance you're trying to find is the diameter of the sun. For purposes of calculation, use only the two uppermost triangles as shown here:

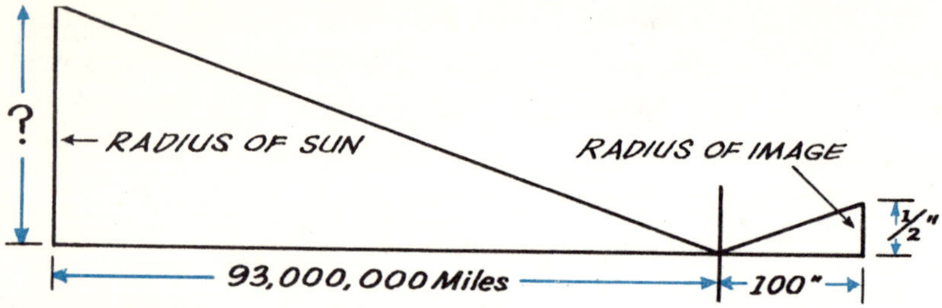

You realize, of course, that the larger triangle is millions of times larger than the smaller triangle. In spite of this, the same principles of proportions apply. For purposes of illustration, assume the distance from the pinhole to the cardboard was 100 inches. *This will not be the distance you use when you work the problem. Use the distance you find by measurement.* 100 inches will be the horizontal line of the smaller triangle. The vertical line will be a half inch or half the diameter of the image of the sun. You can now find the *radius* of the sun by putting this into a proportion formula:

$$\frac{\text{Radius of sun}}{\text{93,000,000 miles}} = \frac{\text{½ inch}}{\text{100 inches}}$$

This proportion could be expressed:

$$\frac{\text{Radius of sun}}{93{,}000{,}000} = \frac{1}{200}$$

By cross multiplication:

$$200 \times \text{Radius} = 1 \times 93{,}000{,}000$$

$$\text{Radius} = \frac{1 \times 93{,}000{,}000}{200}$$

$$\text{Radius} = 465{,}000 \text{ miles}$$

And because the diameter is twice the radius,

$$\text{Diameter} = 930{,}000$$

REMEMBER THIS IS NOT THE CORRECT ANSWER, BECAUSE IT IS BASED ON AN ASSUMED DISTANCE OF 100 INCHES INSTEAD OF THE ACTUAL DISTANCE. THE DIAMETER OF THE SUN IS NOT 930,000 MILES.

When you have figured out the correct diameter of the sun according to the formula, check it in any standard reference book. Astronomers, when they were originally figuring out the diameter of the sun, could not go to a reference book and look it up. They had to check it and check it again in order to make sure their figures were correct.

Astronomers have other ways to measure the distance to the planets and the stars. One way is shown in here:

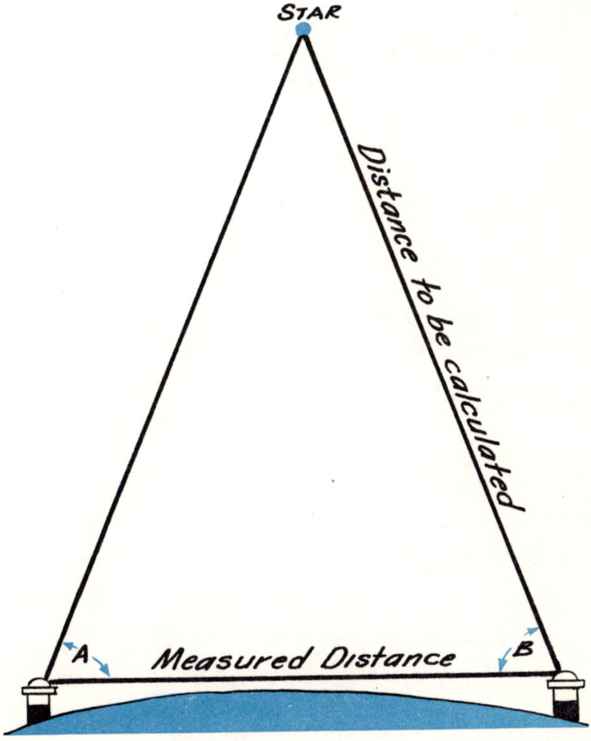

If two astronomers observe a planet at the same time, they can measure the angle indicated as A and B and measure the distance between telescopes. This gives them enough information to be able to calculate the distance to the planet. The greater the distance between them is, the more accurate their calculations can be. You might think that the maximum distance they can get from each other is the diameter of the earth, or 8000 miles. This seemingly long distance is not nearly enough for astronomers to use in calculating the distance to the stars. Instead of the diameter of the earth, they sight the star at exactly six-month intervals. This means the earth has traveled halfway around the sun . . . and the astronomers can use the diameter of the earth's orbit, or 186,000,000 miles!

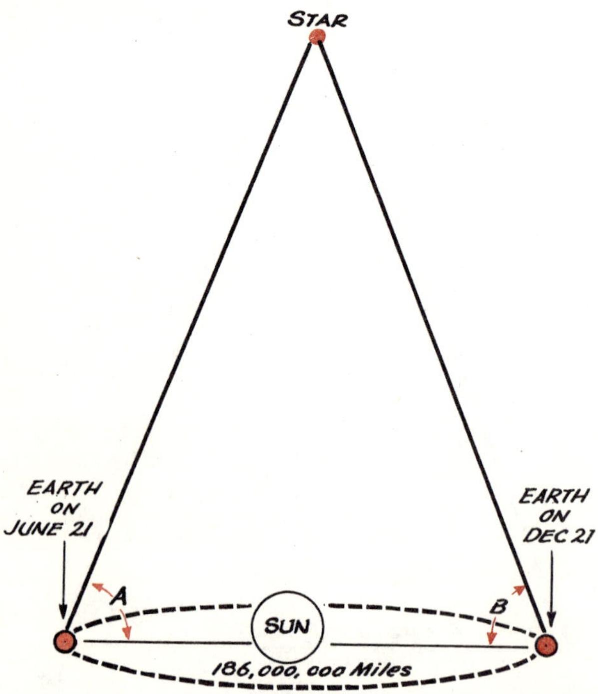

With this distance and the proper angles, astronomers can measure with accuracy a star that is about 1,761,000,000,000,000 miles away! Most of the stars in the sky are still farther away than that, and astronomers use other systems to figure out their diameters and distances. One such system involves comparing the period of dimming and brightening of the light from a star with its general brightness or luminosity. The fact that there is a relationship between these two effects and the distance of the star from the earth was discovered by accident in 1912 by Henrietta Leavitt at the Harvard Observatory. Using this system, astronomers measured stars that are as far away as 150,000 light years. This is 150,000 times 6,000,000,000,000 miles away!

Now that you've solved the same kind of problem that an astronomer does, you can see why an astronomer must be well trained in mathematics.

In order to gather information, an astronomer uses telescopes and many other instruments. Sometimes he uses a special camera to photograph the heavens. The telescope and the camera are driven by motors to automatically follow the stars as they appear to move across the heavens. In reality the telescope and camera move to compensate for the rotation of the earth. With the desired stars remaining for so long in the same position in the "eyepiece" of the telescope, extremely long camera exposures can be made. Many stars become visible on the photographic film that could not be seen by looking through the telescope.

The planet Pluto was discovered by an astronomer working with photographs taken through a telescope. Percival Lowell, an American astronomer, who worked at an observatory in Arizona, had calculated the orbit of the planet Neptune and noted that the actual orbit of the planet did not correspond. He did some further calculating and finally predicted an un-

An astronomer making observations in the prime focus cage of the Hale telescope at the Mount Wilson Observatory. The Hale telescope has a reflecting mirror that is more than 16 feet (200 inches) in diameter. (Photograph courtesy of Mount Wilson and Palomar Observatories.)

discovered planet as the cause of Neptune's strange behavior. For years astronomers tried to find the new planet. On January 21, 1930, C. W. Tombaugh was carefully going over photographs of the section of the sky where the planet should be. He found a point of light that indicated there was an object that fit the predicted orbit and speed. He confirmed his discovery by following the path of the object for several weeks. The finding of the planet Pluto was possible only because of the highly developed skills of scientists working with telescopes, photography, and mathematics.

Often an astronomer examines the light from a star through an instrument called a spectroscope. This is an instrument that breaks up light into colors. Each chemical element gives off light of a characteristic color when it burns. By examining the color of the light from a star, the astronomer can learn something about its chemical composition, atmosphere, speed, etc.

Recently, astronomers have been examining the skies with radio telescopes. These instruments gather invisible radiation from the sky because of their specially designed antennae. Astronomers are finding many new facts about the heavens with their radio telescopes.

Astronomers are seeking the answers we all ask about the heavens. They have given us facts that help us understand the past, present, and future of the universe, as well as how man fits into it. They are working on the answers to such questions as: What is the age of the universe? Are there other suns and planets like ours? How are stars formed and what happens to them? Astronomers are coming closer to the answers to these questions and it's a good thing too. We all want to know the answers out of curiosity, but the astronomer's findings will take on even more importance as we get closer and closer to moving around in outer space ourselves.

2. Living-room Zoo

Zoologist

If you've ever been to the zoo, you know how fascinating it is to watch the animals. Why not do the same thing at home? You not only will be able to watch animals, but feed and study them whenever you have a few moments to spare. You'll find before long you'll be spending hours studying them. The animals you're going to study you'll have to hunt down and catch yourself. Then you'll have to have cages, food, and water. You don't have to go on an African safari to capture your animals, because they're underfoot most of the time. They're ants! It's amazing how easy it is to make an ant zoo.

Here's a list of the materials you'll need:

2	small plastic boxes	Roll of cloth tape
1	larger plastic box	Red cellophane
5	1-inch pieces of ⅛-inch threaded pipe	Cardboard
8	⅛-inch lock nuts	Candle
6″	⅜-inch rubber tubing	Pliers

The plastic boxes are sold in most hardware stores. They're used to store small parts such as fishing hooks, screws, and bolts. The two smaller boxes need not have partitions, but make sure the covers fit tightly. Ants will explore every corner of their new homes, and it's amazing how small a hole they can crawl out of.

The larger box should have several partitions in it, as well as having a tight-fitting cover. When put together as explained later on, they will look like this:

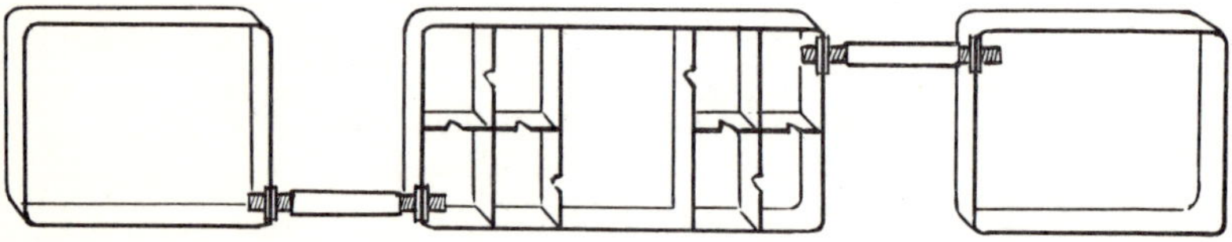

Other kinds of boxes will work just as well. See that they have close-fitting covers and that you can see through them, if not through the top, then through the sides. Some of the plastic boxes used for storing leftover foods in the refrigerator will work nicely. Some foods come in plastic jars and boxes that will work well also. Glass dishes or jars could be used, but, as you will see, they might be hard to connect. If they're the only kind you can get, you might drill holes in the metal tops or fashion wooden covers.

The threaded pipe is sold in hardware stores most often for use in lamps. It's the same pipe that's used in making the microscope described in Chapter Four. It's called "one eighth-inch pipe" but actually has an outside diameter of three eighths of an inch. The rubber tubing fits around it snugly.

The three eighth-inch rubber tubing, the one eighth-inch lock nuts, and the roll of cloth tape are all carried by most hardware stores.

With the pair of pliers, hold one of the pieces of threaded pipe in a candle flame until it is hot enough to melt a hole in the side of the plastic box. It's surprising how little heat is necessary to do this. Pull the rod back and forth as it cools in order to make a loose-fitting hole.

Continue to melt holes through the boxes so they fit together as indicated in the first illustration in this chapter. Insert the four pieces of one eighth-inch threaded pipe, screw on the lock nuts until they are tight, as shown here:

HEAT

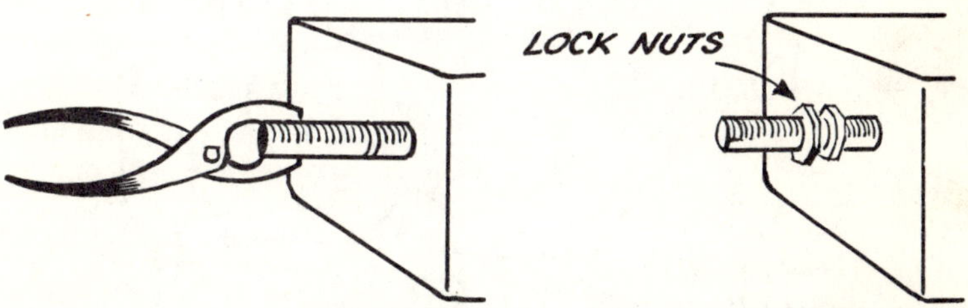

PIERCE

INSERT

If the large box top is held on with hinges, cut them off with a hack saw. With the saw blade or a triangular file, cut notches in the partitions as indicated in the following illustration. If there are partitions in the smaller boxes, you should cut or file notches in them also.

Cut a piece of light-colored cardboard to fit the bottom of the box, and hold it in position as you cover the bottom and sides of the box with the cloth tape. The cardboard provides a light-colored background against which it will be easier to watch the ants. The cloth tape keeps out the light. Ants like to live in the dark.

The reason for cutting off the hinges is to make the inside of the box as dark as possible. With the hinges on, it's difficult to get the tape on so no light leaks in. The top of the large box is seldom removed, so it can be taped in place. Keep most of the top clear so you can look into each one of the chambers easily.

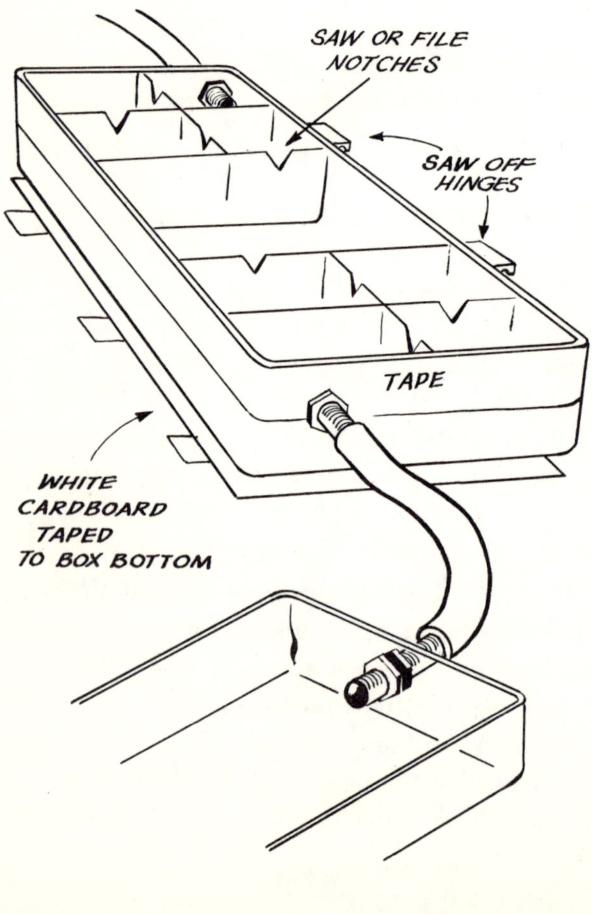

The boxes can now be fastened together with rubber tubing.

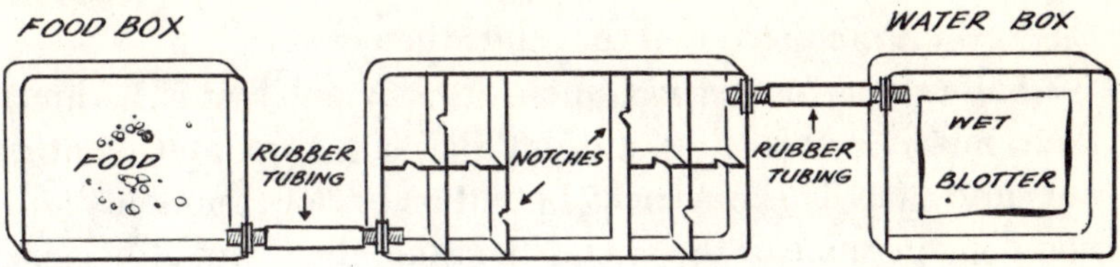

The next step is to find some ants. The easiest way is to turn your cage into a trap. If you're lucky, you may capture a whole colony, queen and all! To get a good supply of ants, you'll have to make the inside of your cage more attractive to the ants than the place they're living in. To do this, you must supply food in a damp, dark place.

Into the left box put a blotter with a few drops of water on it. Add some water to syrup and put some of this mixture on a piece of waxed paper. Put this in the left box also. Close the box and wrap it in heavy paper so that it's dark inside. Open the right box and put a drop of the diluted syrup at the opening of the pipe. Put this trap where you've seen ants. Leave it undisturbed at least overnight. In the morning you may have a whole colony of ants in your cage. If not, wait a few days, checking frequently. When you think you have a a good supply of ants, put some of the diluted syrup on a piece of waxed paper in the right box and close it tight. Cover the center box with cardboard. Now unwrap the left box and see what you've caught. The ants will move into the center box, their future home, because it's dark.

Once you have trapped a colony, put food in the right box and keep a supply of water on the blotter or on a sponge in the left box. Ants need an atmosphere of high humidity. The water is separated from the rest of the cage because ants like their humidity to be just so. With the water at one end of the series of cages and the food at the other, the moisture content of the compartments in between will vary. The ants can find the compartment that suits them just right.

Another way of getting ants for your ant zoo is to hunt them in the "wild." Find an "anthill," dig it up, and put the ants into a jar. This is a tricky operation because, not only will you damage some of the ants in handling them, but also every time you try to put one into the jar, three seem to escape. An easier way is to put ants and soil into a large screw-top jar. The jar top has been prepared as indicated here:

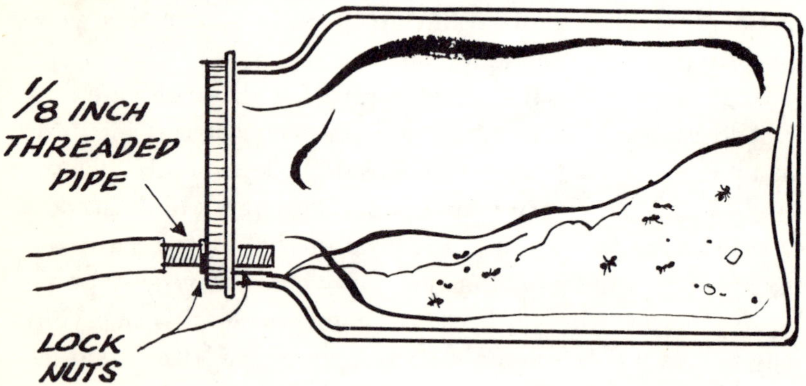

The piece of threaded pipe is attached to the jar top in the same way as it was to the plastic boxes. The hole in the metal jar top will have to be drilled or punched, of course. The hole in the end of the pipe is plugged until you are ready to transfer the ants to the cage. Then the rubber tubing is disconnected from the food box and connected to the tubing in the jar cover. Soon the earth in the jar will dry out and the ants will move into the cage. This may not work with every type of ant, but most of the common variety will move in of their own accord.

Collect as many different kinds of ants as you can from the hill or nest. Look for eggs, larvae, pupae, workers, males, and, if possible, a queen. The ones you find may not look exactly like those pictured below, but collect anything that looks like them. Get workers, males, and queens if possible. The wings may not be in evidence at the time you

are collecting, but if you do see an ant with wings, make a special effort to capture it without injuring it. It will be either a male or a queen. The queen is always larger than the workers, sometimes as much as four or five times as large. If you catch a queen, you will be very lucky indeed. With her you may have a colony of ants that will last for years.

Keep the center box covered while not actually observing the ants. Ants like to work in the dark. Because they cannot see red light, you can cover the top of your box with red plastic or cellophane and see fairly well what is going on in each of the compartments. Some light other than red may come thru the cellophane; therefore it's a good idea to add the opaque cardboard between observations. A magnifying glass on a stand is a great aid in getting a close-up view of the activity within the zoo.

Once you have the ants inside the cage, they will require very little care. Try different foods and record what their reaction to each is. You'll find they have very definite tastes. Among the foods that you should try are: bread, syrup, candy, chocolate, hard-boiled egg, cheese, fresh leaves, or dead insects. Do not keep the food in the box too long. Remove it if the ants have not begun to eat it after a day or so. In any case remove it if you notice it is beginning to spoil.

You'll find that ants can be studied for hours at a stretch and day after day. The colonies are highly organized. Some ants gather food. Others feed and care for the young, keep the nest clean, store food, or guard the entrance.

The worker ants, as their name implies, do the work of keeping the nest going. All the workers are females, but if they lay eggs, and they often do, only male ants can develop. Some kinds of ant nests have only one queen, others have several. The queen ant lays most of the eggs. She is the only ant that lays eggs that can develop into workers or other queens.

The workers may all look alike or be quite different in appearance, often with specially developed parts that are used for their particular job in the colony. Ants within a colony do not quarrel with each other. If two ants seem to be fighting, they are probably only having a friendly tussle. They do fight to the death whenever their homes or their young are threatened by other ants or other animals.

You can see from this very brief description that you can spend many hours watching the ants set up housekeeping in your zoo. Watch one particular ant as she goes about her work. In this way you'll learn how each ant contributes to the welfare of the colony.

Ants are part of the group of animals called insects. A scientist who specializes in the study of animals is called a zoologist. So as you get to know more about your ants you can think of yourself as a junior zoologist.

Animals have been very closely associated with man from the beginning of history. Man has been dependent on animals for honey and leather, soap and ham, and thousands of other products. He has had to live with the bites of mosquitoes, the fear of snakes, the invasion of parasites. Man himself is an animal. Zoologists have helped control man's animal enemies and develop man's animal friends.

One of the early problems that confronted the zoologist was the classification of animals. There are about 1,000,000 different kinds of animals. Getting them classified in some logical order was a necessity if scientists were going to study and exchange information about them. You probably think of various animals as belonging to groups which you call snakes, dogs, cats, frogs, birds, insects, fish, or some other word that denotes a group of similar animals. The zoologist uses a similar system for classifying animals but has much more precise definitions and names.

To you a spider may be an insect, but to a zoologist a spider is not an insect at all. Both insects and spiders belong to the same general group of animals. The zoologist calls this large grouping or "phylum," "Arthropoda," (meaning: "joint footed"). Spiders, because of their many differences from insects, the most obvious of which is usually their four pairs of legs, are classified in the class "Arachnida." Insects, on the other hand, have three pairs of legs and belong to the class "Insecta." Most insects are animals with a body divided into three parts: the head, containing the mouth parts and sense

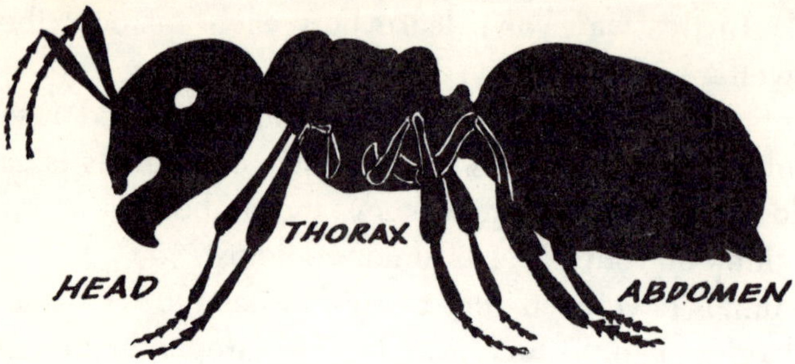

organs; the thorax, with the legs and wings (if any); and the abdomen, in which are located the breathing, digestive, excretory, and reproductive organs. Remember, the spiders usually have four pairs of legs and insects usually three pairs.

The class "Insecta" is further divided into "orders." One of the orders is called "Hymenoptera," and within this order are ants, bees, wasps, sawflies, gall wasps, and others. This order forms one of the largest and most highly developed order of insects. There are more than 118,000 different kinds of insects in this order. They, like other animal classifications, are further classified into sub-orders and super-families and families. Ants belong to the family Formicoidea. Within this family are subfamilies and genera and finally the particular species of ant. You might find it an interesting challenge to track down the scientific name of the species of ant you have captured.

Most reference books will give you information about ants, and there will probably be several books on zoology in your library. Zoology is such a vast field with so many animals and so much to be studied about them that most scientists specialize in some way. You could become an "entomologist"

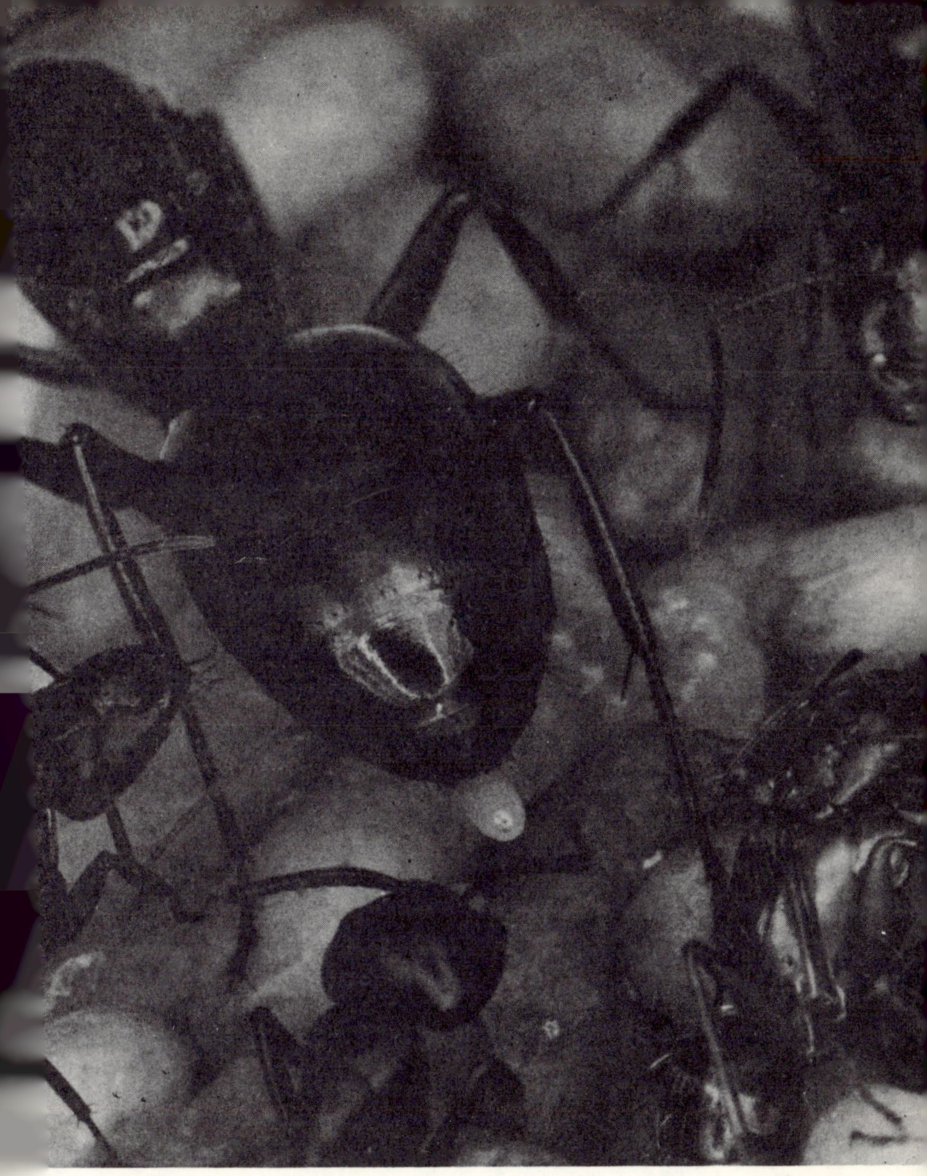

If you capture a queen ant, and your ants have a good supply of food and water, you may be able to observe, through a magnifying glass, the queen ant laying eggs like this. The ants around the queen are special workers that press the rear part of her body to help the new egg arrive. (Lennart Nilsson—Black Star.)

if you continued your study of insects. If you specialized in the study of birds, you would become known as an "ornithologist." Snakes, a "herpetologist," fishes, an "ichthyologist," and so on.

If you confined your investigation to one aspect of the animal world, such as classification of animals, you would be known as a "taxonomist." Other fields and the name of the scientist who works in it are given in the following list.

Science of heredity and breeding	Geneticist
Development of the egg	Embryologist
Bodily functions	Physiologist
Study of intelligence	Psychologist
Environment as it affects animals	Ecologist
Diseases of animals	Pathologist
Chemistry of living things	Biochemist
Physics of living things	Biophysicist

As you probably noticed, many of the fields that are subdivisions of zoology are closely related to medicine. This is not surprising when you remember that man is an animal. Because of this close relationship to medicine, many zoologists are employed by hospitals and medical centers.

What do you think will happen to your ants as time goes on? Unless you have captured a queen, the colony will eventually die out. This may take as long as six months, however, so you will have a good chance to observe them before you'll have to get a new supply. During that time you can be sure you'll spend many fascinated hours watching the endless activity of the phylum Arthropoda, class Insecta, order Hymenoptera, family Formicoidea!

3. The Invisible Seeds

Microbiologist

Did you know you can plant invisible "seeds" and then see them grow into a garden? You might logically ask: "How can you plant the seeds if you can't see them?" It's really very easy. As you watch the seeds grow and get to know more about them, you'll be a microbiologist.

Here's what you need to grow your invisible "seeds":

Can of tomato soup
6 small custard cups
6 sheets of clear plastic (Saran Wrap)
6 rubber bands
Tablespoon
Kitchen tongs
2 saucepans

The reason you need six custard cups is because there are six different gardens to grow. If you don't have six dishes, plant as many gardens as you can. Any small dishes will do, as long as they can withstand boiling in water.

Here's what you do: Open the can of tomato soup, pour it into one of the saucepans, add half a can of water, cover, and bring to a boil. Let the soup simmer for at least twenty minutes.

The clear plastic is the type used to wrap sandwiches and cover food dishes. Cut the plastic into sheets at least three times the size of the dishes. You're going to boil the sheets and they shrink!

Into the second saucepan put the dishes, plastic sheets, kitchen tongs, and tablespoon. Cover with water, bring to a boil, and continue boiling for at least twenty minutes.

While the saucepans are boiling, prepare a place to keep the dishes. A tray or piece of board is handy. Once you have the dishes filled, you can carry them to where you can conveniently observe them.

When the twenty minutes are up, carefully remove the tongs with a fork or spoon touching only the handle end of the tongs. Let them cool enough for you to handle.

What to do with the tongs while you are waiting for them to cool? You don't want to set them down on the table. One answer is given here:

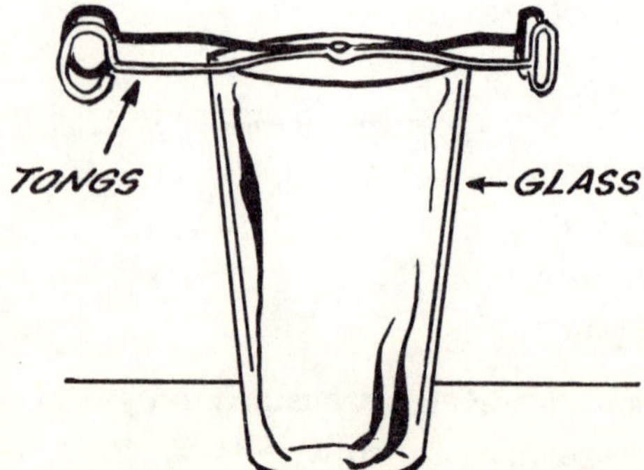

Be especially careful from now on because everything you'll be working with will be very hot. There's another reason for being so careful which you'll find out about later on. With the tongs remove a dish, empty the water in it, and place it on the sink or table. With the tongs, remove the tablespoon, and when it's cool enough to touch, put two tablespoons of tomato soup into the dish. Pick up one of the plastic sheets with the tongs. Let it cool enough so you can handle it by the edges. It will be crumpled up into a ball, but you can open it by pulling gently on the corners. Be sure not to touch the center part. Put the plastic sheet over the dish, smooth it down around the edge, and slip a rubber band around it. The first dish will now look like this:

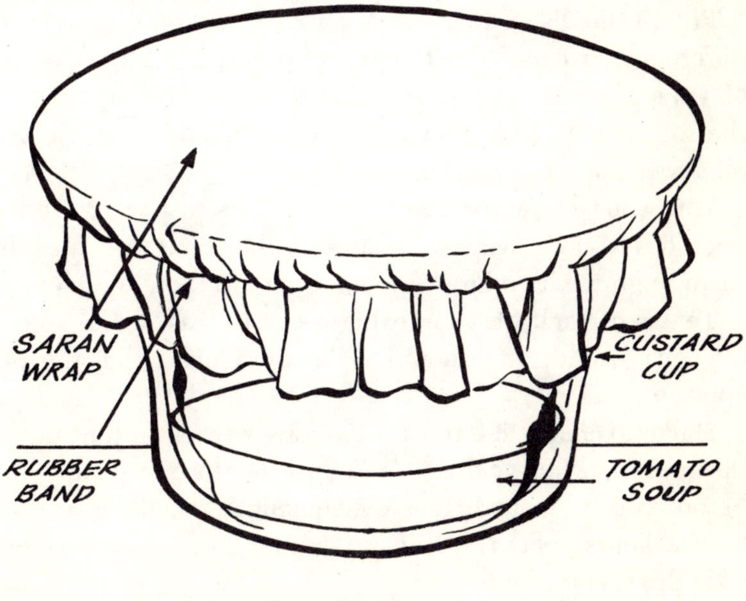

Dish 2. Remove, add two tablespoons of tomato soup and set it aside without covering it. This is to allow the surface of the soup to be exposed to the air in the room. Allow it to re-

main this way for half an hour. Then cover it with a plastic sheet as you did the first dish.

Dish 3. Remove from the saucepan, add two tablespoonsful of tomato soup, and when it's cool enough, touch one of your fingers to the top surface. Cover the dish with the plastic sheet.

Dish 4. Add tomato soup as before but do not cover. Wash your hands thoroughly with soap and water but do not dry your hands with a towel. Touch the surface of the tomato soup with one finger and then cover with the plastic sheet.

Dish 5. After adding tomato soup, rub a finger on the floor, touch the surface, and cover.

Dish 6. Add the tomato soup, sprinkle a few bread crumbs on the surface, and cover.

Put all the dishes in a warm, dark place. You have now six different "gardens." The tomato soup in the dishes is the soil of your gardens. You have planted the "invisible" seeds in the soil of only five. The seeds you've planted are the tiny microbes that are everywhere.

You won't know for three or four days just what kind of microbes you've planted because it will take that long for them to grow into a large enough group for you to see.

To keep track of your experiment just as a real microbiologist would, you should make up a chart like the one opposite.

Make up charts like this for two observations a day, one in the morning when you wake and the other for the evening just before you go to bed. It is convenient if the times are just twelve hours apart. That way you are checking your seeds' growth at regular intervals.

What should you look for? Well, let's start at the beginning of the experiment so you understand why a microbiologist works the way he does.

DATE: TIME: PLACE: AIR TEMP: HOURS PLANTED:		
DISH 1	CONTROL	CONDITIONS:
DISH 2	AIR ½ HR.	
DISH 3	DIRTY FINGER	
DISH 4	CLEAN FINGER	
DISH 5	FLOOR	
DISH 6	BREAD CRUMBS	

Remember you boiled everything for at least twenty minutes? This was to kill all the microbes that were already on the equipment and in the soup. You then handled the equipment very carefully; this was to prevent the microbes on your hands from getting into the soup. You quickly covered the first dish, and if no microbes were alive inside and you allow none to get inside . . . nothing should grow in dish 1. This is your "control." It gives you a dish to compare with the others. If something should develop inside dish 1, it means either the heat didn't destroy all the microbes or some got inside before you covered it.

You exposed dish 2 to the air for half an hour. The microbes that were in the air had a good opportunity to fall on the soup. Thousands of different organisms could be in that dish, but the microbes that will grow large enough for you to see without a microscope will probably be bacteria or molds.

Bacteria and molds are plants, but not like the green plants you're familiar with. They contain no chlorophyll or food-producing chemicals. They must get their food from other plant or animal material. The tomato soup is the food you have provided for them.

Bacteria are tiny plants that are almost everywhere except in sterilized places—such as dish 1. They're so tiny you can't see them except with a microscope under the highest magnification. They're only about 1/400,000 of an inch long!

If one lands on the tomato soup, the chances are that the conditions will not be favorable for its growth. Most bacteria need temperatures higher than room temperature to grow. If one bacterium did manage to grow, it would reproduce by dividing and becoming two bacteria in a half hour or so. These two could become four in an hour. At this rate there will be millions and millions of bacteria at your first observation twelve hours later. You still may not be able to see them for a few days. When there are enough of them to see, they'll appear as a spot on the surface of the tomato soup. If you see several spots, each one probably started from one bacterium. Record their appearance every twelve hours.

Most of the spots on the surface of the tomato soup will be molds. You've probably seen molds on bread, jelly, cheese, and other foods. Molds are plants that do not produce their own food. In this way they are like bacteria. However, there are many differences. One important difference is the manner in which they reproduce. Bacteria, you remember, reproduced by dividing into two. Molds, on the other hand, reproduce by seed-like "spores." It is these spores that can fall onto the surface of the soup. Some spores are so small they can "float" in the air. They're not quite as small as bacteria, but are too small to be seen without a microscope. The plants that grow from them are easily seen. There are many different kinds of molds. You will probably get at least three different colors of mold plants. A very common mold is one microbiologists call "*Mucor mucedo.*"

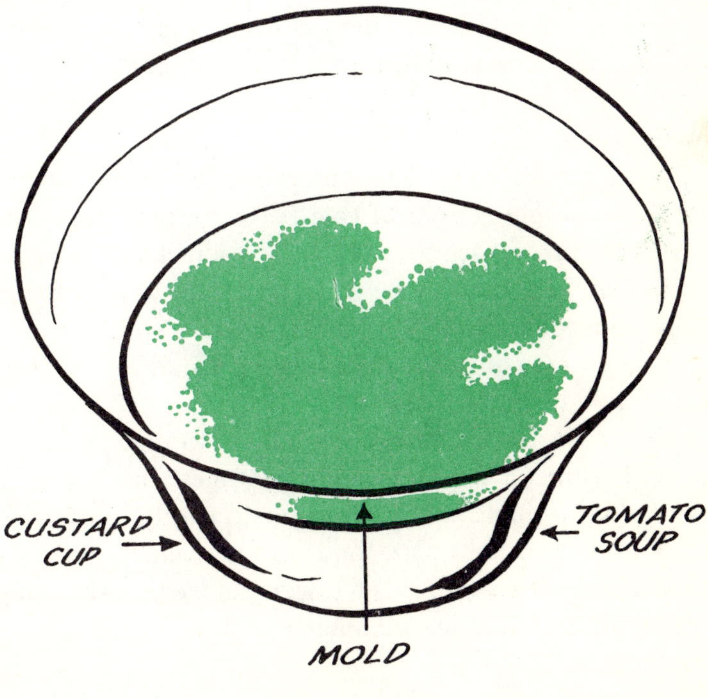

In order to find out what "seeds" you've planted, you'll have to ask someone who knows about bacteria and molds or, better yet, go to the library and find books on them. You'll have to have a very powerful microscope to identify bacteria but some of the molds can be studied with a magnifying glass.

Dish 3. Dish 3 was the dish you touched with your finger. Dish 4 you touched with a specially washed finger. The third dish will probably have much more mold growing in it than dish 4, where you made a special effort to get your hands really clean. Even with all your scrubbing, there may be colonies growing in dish 4. Your hands were clean enough to eat dinner, perhaps, but they still had thousands of bacteria and mold "spores" on them.

Dish 5. You rubbed your finger on the floor and then touched it to the surface of the soup. The floor, of course, had a good supply of bacteria and molds because they are brought in on your shoes. This should be one of the most successful of your gardens.

Dish 6. You sprinkled a few crumbs of bread into the sixth dish. It, too, should be one of your better gardens, because the things that you eat are covered with various kinds of bacteria and molds. Some of the organisms growing in the last dish could have come from your hands, because you touched the bread as you put it in the dish, but the bread also contained many different kinds of microbes.

You can now investigate many "clean" things. You can devise various methods of collecting microbes from around the house. For instance, a sterile gauze pad can be used to wipe across a "drip-dried" plate and another gauze pad wiped across a plate dried with a dish towel. Put each gauze pad into sterile soup to find out which has the highest number of microbes on it.

You can also sterilize things by baking them for at least an hour. The time is important, because some bacteria could survive a few minutes of heating. Baking and boiling kill the microbes by heating them until they are literally "cooked." Some chemicals such as iodine and alcohol kill bacteria and molds by poisoning them.

The first man to seriously study these tiny plants was Louis Pasteur. If you haven't read about the fascinating discoveries he made in the 1800s, by all means do so while your own microbes are growing. You will most likely be dealing with some of the same kinds of microbes Pasteur dealt with!

In 1928, in a hospital in London, a bacteriologist (one kind of microbiologist) named Alexander Fleming was studying a group of bacteria or germs which caused infection. He was doing this by growing them in small dishes just as you did. Accidentally, some of the mold spores got into one dish. When Fleming investigated he discovered that wherever the mold was growing the bacteria were destroyed. Fleming guessed that the bacteria were killed by the mold or some chemical produced by the mold. He then cultivated the mold and found it produced a yellow material that restricted the growth of some bacteria. He called the yellow material "penicillin" because the name of the mold was "*Penicillium.*" This is a common mold, one that could be growing in one of your dishes.

The accurate records you're keeping of your "gardens" tell exactly what is going on. This is what Fleming did, and it was because of this that sometime later Howard Florey, a professor at Oxford University, was able to repeat Fleming's experiments and get the same results.

Florey tried to remove the penicillin from the mold. He got some that was almost pure but not pure enough. He was joined by Ernst Chain and together they finally produced pure white crystals of penicillin. This was the form in which

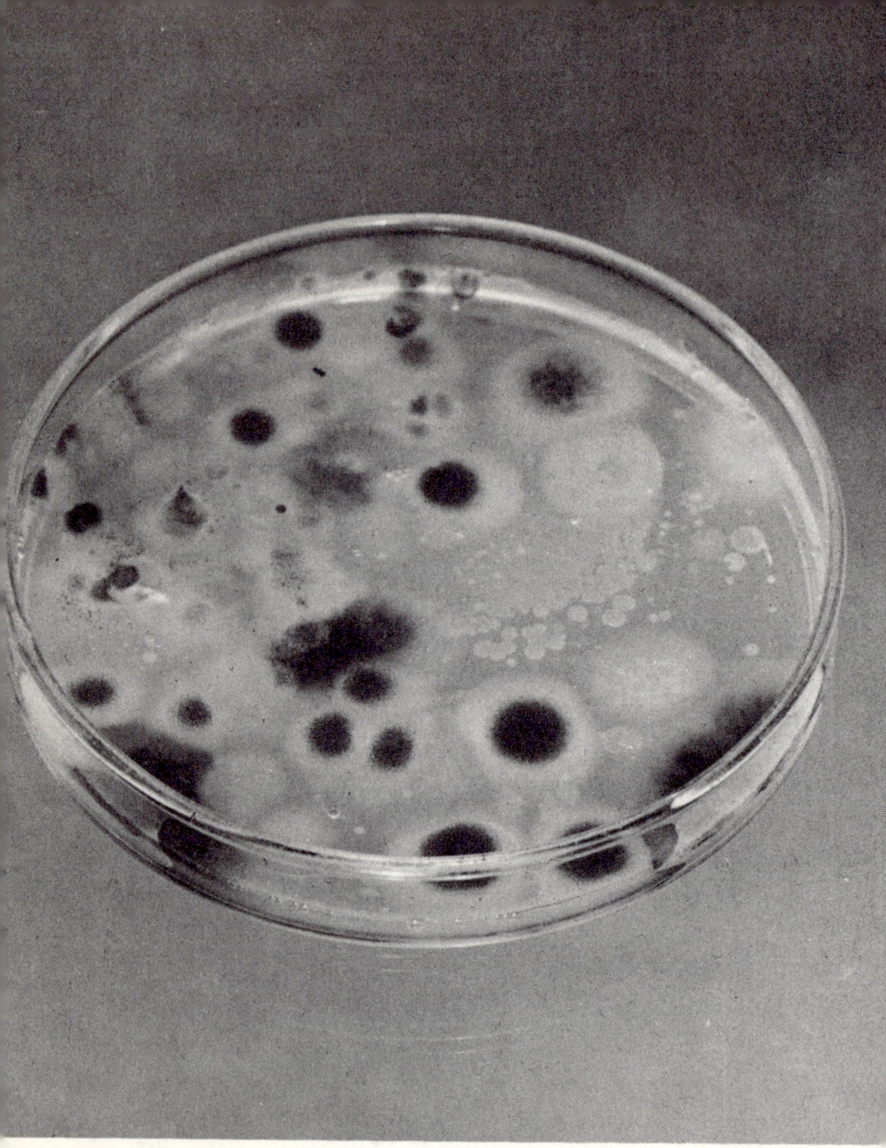

The invisible "seeds" you plant may look like those in this dish that was prepared by a real microbiologist. These "spots" have developed from the molds and bacteria that were in samples of soil placed in the dish with sterile food and water. The microbiologist separates the growths for further study. (Photograph courtesy of Chas. Pfizer & Co., Inc.)

it was needed to finally become such a good germ killer. Fleming, Florey, and Chain have received many honors for their development of penicillin.

Other microbiologists helped perfect the process whereby penicillin could be extracted from the actual mold. Still other microbiologists worked out methods of producing the drug on a large scale. Today the search for materials similar to penicillin continues. Microbiologists, of course, are leaders in this search for new "antibiotics."

The microbiologist works in many fields besides medicine. His work is around you everywhere. Here are some of the things that may be around the house that have been prepared with the help of a microbiologist: vinegar, cheese, bread, rolls, wine, vitamins, yogurt, paint spray, beer, cosmetics, plastics, animal foods, blueprints, paints, rust remover, candy . . . in fact, almost anywhere you look you'll find the work of a microbiologist. Certainly everywhere you look you know by now you could find the "invisible seeds" he knows so well.

4. Crystal Garden

Physical Chemist

How would you like to have a whole dishful of crystals that "grow" like plants and look like coral? To grow such a dishful of crystals is relatively easy. Here's what you'll need:

Charcoal briquettes
Water
Table salt
Laundry bluing
Ammonia
Small bowl
Large plate
Tablespoon
Notebook
Pencil

The charcoal briquettes are the kind that are sold in hardware and grocery stores for making "charcoal broiled" steaks and other foods. Put several pieces in the bowl and prepare the following solution:

¼ cup water
¼ cup laundry bluing
¼ cup table salt
Tablespoon ammonia

Stir well and pour over the briquettes in the bowl. Place the bowl on the large plate, and put the whole thing someplace where it will be undisturbed. Be sure the pieces of charcoal stick up out of the solution. Try putting drops of mercurochrome, colored inks, or food coloring at various places on the surface of the briquettes.

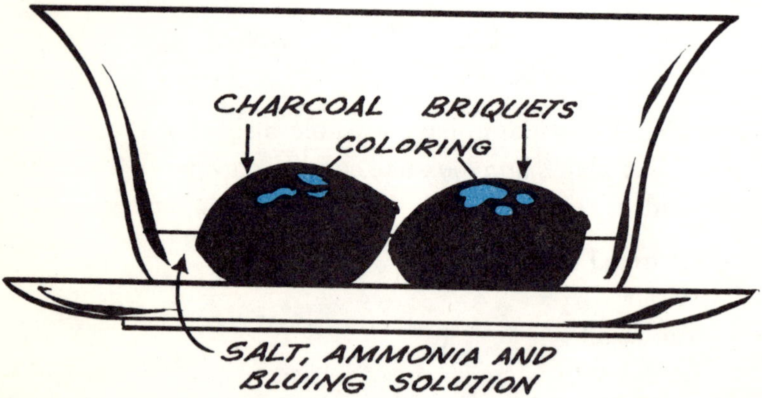

Keep accurate notes on the exact amounts of each material, the date, time, temperature, and the appearance of the materials in the bowl. Make observations and record conditions every few hours on a regular basis if possible.

How soon the crystals will begin to appear depends on the temperature and humidity in the room. It's possible that in an hour you will begin to see crystals forming on top of the charcoal. Before long, crystals will be growing all over the briquettes, on the sides of the dish, and even over the edge of the dish and down onto the plate. Except where they are colored by the inks or food coloring, the crystals will be pure white with a snow-like texture that's beautiful and delicate.

"Coral-like crystals." This is the kind of crystals that will form on the pieces of charcoal in a short while. (Photograph courtesy of Prism Productions, Inc.)

As you observed the growing crystals, you were beginning to look for the things a physical chemist would. He is a scientist who studies how the chemical parts of materials are physically put together. He is working in both the fields of chemistry and physics.

A physical chemist studies how many things are put together, but the special branch of his work called "crystallography" (the study of crystals) is the field you have started to investigate. A physical chemist would explain your bowl of crystals like this:

You mixed together a solution of various salts in water and poured this over the charcoal briquettes. The briquettes are full of tiny empty spaces inside, and the water was drawn into the tiny spaces. Soon the water was evaporating from the surface of the liquid in the bowl. Whenever it evaporated near something solid like the surface of the briquette or the side of the bowl, the salt that was in the water was left behind to form a tiny crystal on the solid surface. These are the first crystals you saw.

These crystals are very unusual. They, too, have tiny spaces inside of them and the water from below is drawn up and out to the ends of the crystal. Here the water evaporates and the salts left behind form a new part on the old crystal. This "growing" continues until all the water has evaporated and the whole bowl and plate are covered with beautiful crystals.

Unfortunately, the crystals cannot be handled because they are very fragile. It's best not to move the dish at all, because the slightest jar can make them collapse.

The crystals in the bowl are made up of complex salts of ammonia, sodium chloride, and bluing. Other materials form crystals, too, but not like those in the bowl. In fact, salts, some acids, and many things formed by living matter form crystals that have individual shapes. The crystals may be so small you have to examine them with a microscope, or they may be large enough to measure with a foot ruler. All crystals of the same material have the same shape regardless of the size of the crystal.

X rays have been especially helpful to the physical chemists in their examination of crystals. X-ray photographs have helped them develop the theory that each crystal takes its characteristic form because of its unique combination of atoms and molecules.

Crystallographers have classified the various shapes and found that there are thirty-two basic types. The thirty-two types can be grouped into six major systems, some of which are illustrated here:

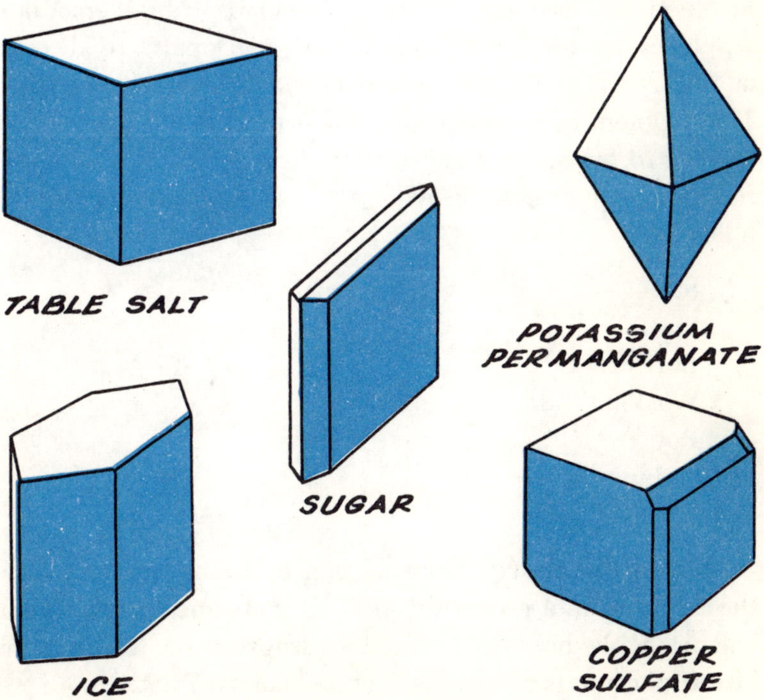

The easiest one for you to examine is the cubic form of table salt. You may have to examine the salt "grain" with a magnifying glass in order to see that it is a tiny cube. Some salt found on grocery shelves today is not cubic in form, probably because of chemicals that have been added to make the salt flow freely in damp weather.

Hundreds and hundreds of different kinds of crystals are formed in nature. A snowflake is a crystal. So are most of the minerals found in the ground. A long time ago the materials in the earth were so hot they melted. When this rock solution cooled, the minerals crystallized. This is how quartz, feldspar, and mica crystals were formed in a layer of granite.

A gem is a crystal also. It happens to be so beautiful that people like to wear it. Gems for jewelry are also hard and can be cut and shaped and polished. Diamonds, rubies, emeralds, sapphires, agates, amethysts, garnets, and opals are all crystals. Why not make a crystal that will be more like a gem? The problem of growing a "gem" crystal is not getting the crystals to grow, but to keep them from growing. The materials you can use to grow "gem" crystals are any of the following:

Salt
Sugar
Boric acid
Washing soda
Borax
Sodium bicarbonate
Alum

Most of the above chemicals can be found either around the house or in a grocery store. Other chemicals, which you will probably have to get at the drugstore, will also grow beautiful crystals of various sizes and colors. Try these:

Potassium chromate
Potassium dichromate
Potassium permanganate
Ferrous sulphate
Manganese sulphate
Nickel sulphate
Copper sulphate
Cobalt chloride
Zinc sulphate
Sodium thiosulphate (photographer's hypo)
Magnesium sulphate (Epsom salts)
Ammonium chloride

A word of caution; some of these chemicals are poisonous if taken internally, so handle them with care and wash your hands thoroughly when you have finished working with them.

Here's how to go about growing your gem of a crystal. Dissolve only one of the chemicals in a cup of very hot water. Continue to add the chemical until no more will dissolve. Put the solution aside until it is cool. This is now a "saturated" solution.

Pour a small quantity of the solution onto a plate, and put the remainder of it in a jar with a wide mouth and a tight-fitting screw top. As the water evaporates from the plate, crystals will form just as before. Look among these for a tiny crystal that has a regular shape. Carefully remove it with a tweezers, toothpick, or small splinter of wood, and put it in the jar with the remaining saturated solution. Do not put the top on the jar. As the solution evaporates, the crystal will grow larger and larger. The reason for removing a "seed" crystal from the rest of the crystals is to prevent it from being deformed. If other crystals begin to grow in the solution in the jar, or the seed crystal gets misshapen, put the cover on the jar to stop evaporation. The smaller crystals will gradually disappear and the larger one will grow faster and better. The best way to form almost perfect crystals is to grow the crystals slowly. This can take anywhere from a few days to several weeks. Do not touch the jar or move it about because this will cause tiny splinters of crystals to form, and each one of these will start growing a new crystal.

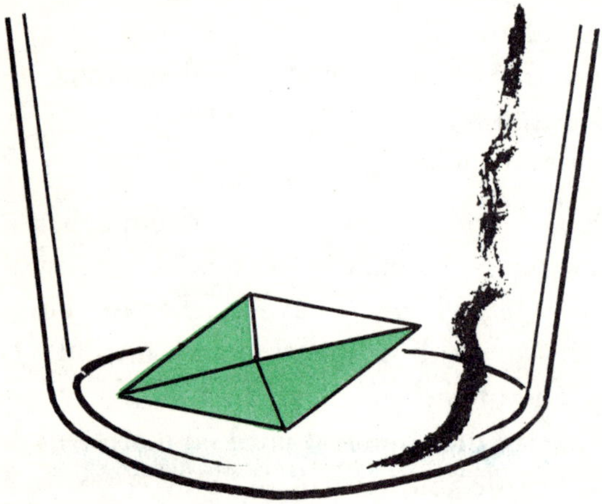

Change the position of the growing crystal several times a day, if possible, so that all the faces will be exposed to the liquid. The face of the crystal that touches the bottom of the jar does not increase in size. If you notice that the edges or the corners of the crystal are getting rounded, pour off the liquid and put in fresh saturated solution. Otherwise the crystal will not grow properly.

Continue to grow the crystal until it gets as large as you would like it to be. Then remove it carefully and wash it with the proper liquid. This may be alcohol, carbon tetrachloride, benzine, cleaning fluid, or some such fluid. Do not wash it off with water, because the crystal will be dissolved. Try washing another crystal of the same material first to find out which is the best solution to wash that particular crystal in. If you leave some kinds of crystals in the air, they can lose water that is contained within them. This will make them dull appearing. Those particular crystals should be left in their cleaning fluid.

You can grow a whole string of crystals that will look something like jewels on a necklace. And what's more surprising, you can eat them because they're made of sugar. Your materials are:

Sugar	Pan	String
Water	Spoon	Paper clips
Cup	Pencil	Glass jar

Pour a cup of boiling water into a dish and add one and three quarters cups of clean granulated sugar. Stir until the sugar is completely dissolved. Put the solution aside to cool. While it's cooling, prepare a glass jar as shown here:

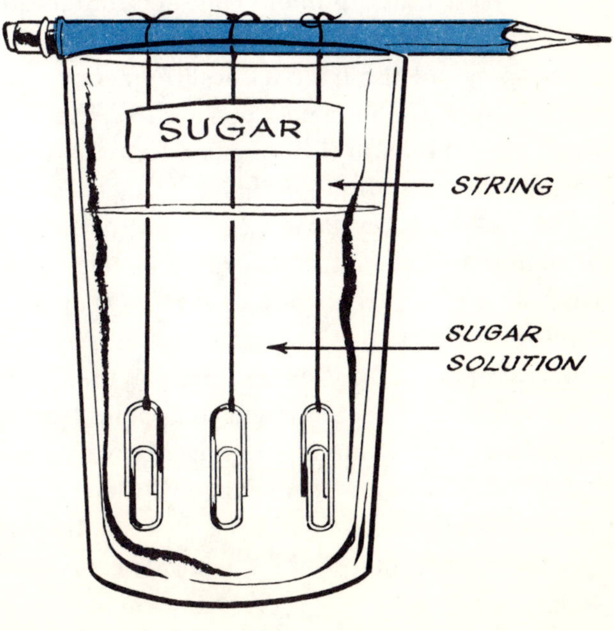

Use clean string, and wash the paper clips thoroughly. Other weights can be used if necessary. Be sure to label the jar so you won't mistake it for another. When the sugar solution is cool, pour it into the jar where it will be undisturbed. Small crystals may form on the string in a few hours. The crystals may grow to be as large as half an inch on each side. You can preserve them if you like by taking them out of the solution and keeping them dry. Guard them, however, because they're good to eat.

To find a good supply of natural crystals is easy. Pick up a handful of sand and look at it closely. The tiny "grains" you see are quartz crystals.

Quartz crystals are very important in radio, television and radar. When they are sliced into plates, with two flat surfaces and then squeezed, one face develops a positive charge and the other a negative. Pierre Curie, a physicist, discovered this effect in 1880. The opposite is also true. If a voltage is applied to a quartz crystal, it expands and contracts, or vibrates at the same frequency as the voltage alternates.

The quartz crystal also has a natural frequency of vibration that depends upon the thickness of the slice. The thinner the slice, the higher the natural frequency. Extremely thin slices have been made that vibrate millions of times a second. This property of the quartz crystal is used to control the radio wave that is broadcast. The fact that a radio or television station broadcasts at a given frequency or certain spot on the dial is due to the thickness of the quartz crystal that is used.

The process of crystallization is sometimes used by chemists to purify chemicals or separate one compound from another. By forming a crystal, some of the impurities in the solution are left behind. They can then redissolve the crystal in pure water and crystallize it a second time. By repeating this process many times, extremely pure chemicals can be obtained.

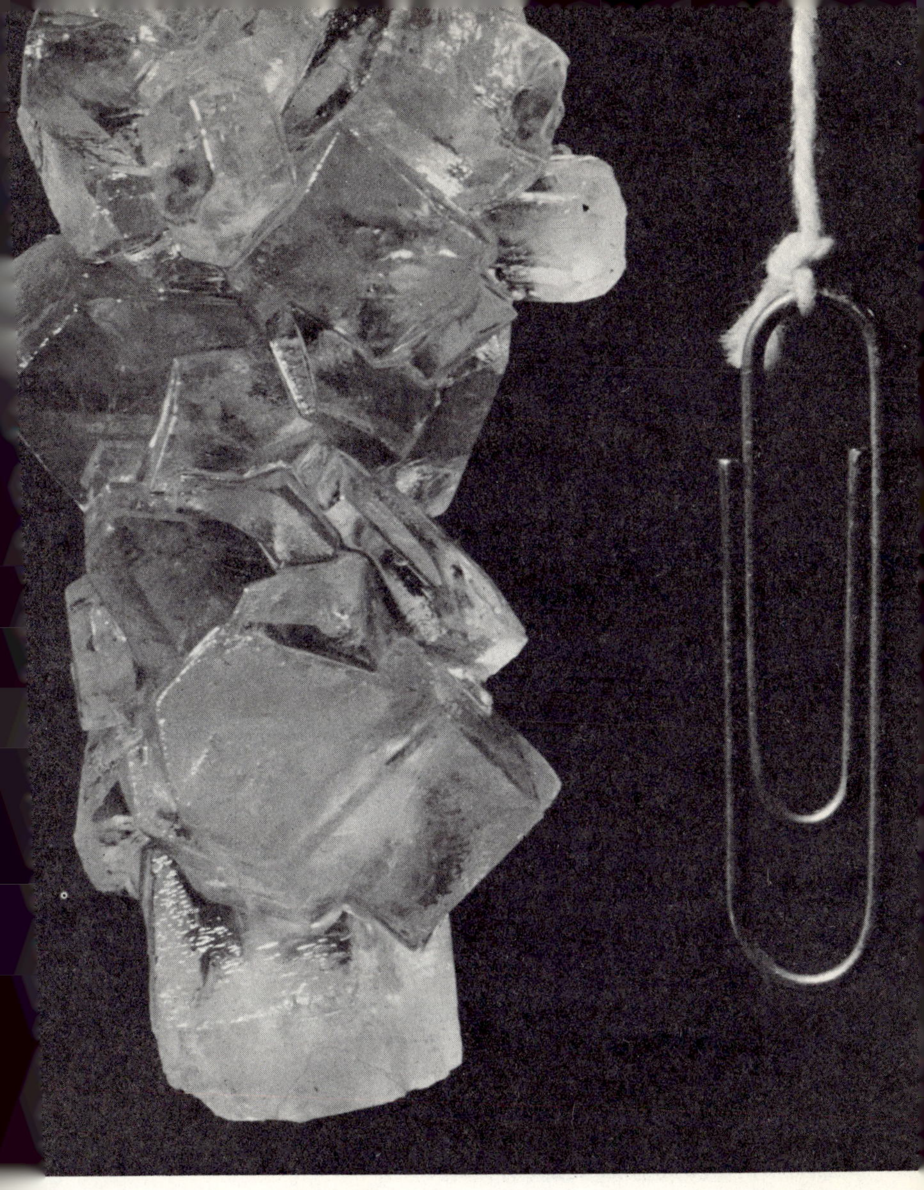

"Crystals on a string." These are sugar crystals that were made on a string in much the same way you can make them by following the instructions. These crystals are sometimes called "rock candy" and the sugar crystals do look like rock crystals. (Photograph courtesy of Prism Productions, Inc.)

Pierre Curie and his wife, Marie, who are so famous for their discovery of radium, used this method to get a thimbleful of radium from more than eight tons of raw material. It took them eight months of crystallizing to do it.

Germanium at one time was a little-used crystal. Further investigation by physical chemists led to the development of the transistor. Almost absolutely pure germanium crystals are grown for use in transistors.

Every piece of metal is a mass of tiny crystals. Researchers at the General Electric Research Laboratory are growing pure crystals of iron that prove to be 150 times as strong as ordinary iron crystals and four times as strong as the strongest steel wire.

Out of such research will come the new materials of tomorrow. An important scientist in the future will be the physical chemist . . . the man who works with the physics of chemistry and the chemistry of physics.

5. Glue from Milk

Organic Chemist

Next time you drink a glass of milk, remember that you could have made glue out of it. An organic chemist would be able to turn milk into, not only glue, but into paint, buttons, and other things as well. To see how this is possible, why not become an organic chemist and actually make glue out of milk?

MATERIALS:
Skim milk
Vinegar
Glass or enameled pan
Sodium bicarbonate (or other base)
Tablespoon
Container for glue

Put a pint of skim milk and six tablespoons of vinegar into a glass or enameled saucepan and heat slowly, stirring continually. Be sure to use only skim milk. As you stir, you will see the milk gradually curdle or form into small lumps. As soon as it begins to do this, remove it from the heat. Continue stirring until the curdling stops. Pour the curdled milk into a convenient container and let the curdled part of it settle to the bottom.

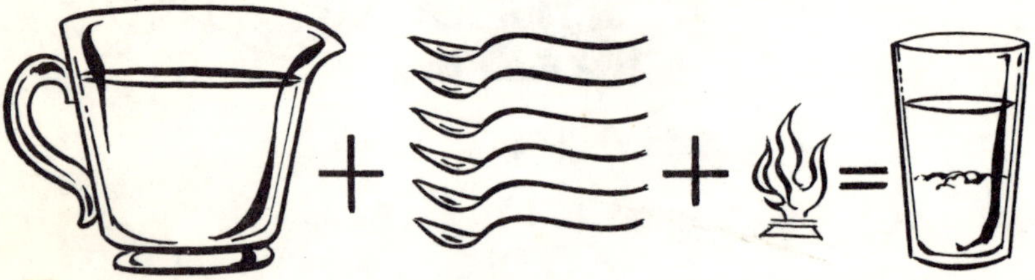

Do you realize what you've made? You've made the curds and whey of Little Miss Muffet fame. The curds have settled to the bottom and the liquid above them is whey. You have artificially "soured" the milk by adding vinegar which the organic chemist would call "dilute acetic acid." A very similar method is used to form the curds that, with a little more processing, become cottage cheese. When the milk sours naturally, the souring and curdling are caused by lactic acid formed by the action of bacteria.

Pour off the whey and let the curds settle. Then pour off the whey again. Continue to do this until the curds are dry. An alternate method is to put them in a strainer to drain thoroughly. Add to them a quarter cup of water and a level tablespoon of sodium bicarbonate or other convenient base, and stir. (For what a base is, see the chapter on the analytical chemist.) As a substitute you could add borax or washing soda. The sodium bicarbonate will cause very fine bubbles to form. This is because of the chemical reaction between the sodium bicarbonate and the vinegar that remains in the curd. You now have glue.

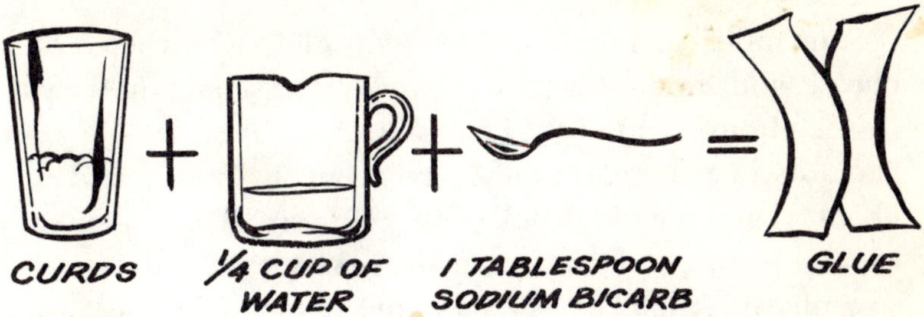

Test your glue by pasting together two pieces of paper. Let it dry thoroughly and you'll find when you try to separate the paper, it tears before the glue gives way. It's called casein glue because casein is another name for the curd.

If you would like to make "fancy" glue of various colors, simply add the food coloring used for cooking. Red, blue, green, and yellow glues are easy to make this way.

GLUE CAN BE MADE IN ALL THESE COLORS

Another way of curdling milk is by adding rennet. Rennet is an organic chemical called an enzyme and comes from the stomach lining of a calf. Milk is curdled with rennet in the process of making cheese. About five quarts of milk are needed to make one pound of cheese. You may have used a similar method of curdling milk at home when you made custard. Next time look for the word "rennet" on the package of custard.

Skim milk is an important by-product left over when cream, cheese, and various other dairy products are made. In the past it was often fed to hogs. Things changed as soon as organic chemists began to find out more about it. They found that about 3 per cent of whole milk is casein. This casein is a protein made up of carbon, hydrogen, oxygen, sulphur, and phosphorus. The chemists found out how to take these components apart and combine them in different ways.

One result of such research is a fiber called "Aralac." It can be blended with rayon, cotton, mohair, wool, or fur in different amounts to produce fabrics of exceptional beauty. Aralac is less expensive than fur or wool, but more expensive than cotton. When added to cotton or wool clothing, it increases the insulating properties and makes the clothing warmer.

Casein can be made into combs, beads, buttons, umbrella handles that look like tortoise shell, amber, ebony, horn, agate, malachite, and other materials. This is one reason why costume jewelry and decorative materials are made of casein. The raw curd is washed, dried and ground, mixed with chemicals, and forced through dies to form rods. These rods are hardened with formaldehyde and dried. Blocks and sheets are made by pressing several rods together and hardening them.

The glue you made is very similar to the white glue in a plastic bottle that's sold in stores. The commercial variety is more refined and has greater holding power than the glue you made. Soft-drink labels are glued to the bottle with casein glue. Casein paints are water-based paints that are inexpensive and come in a great variety of colors. Most water-based paints have casein glue in them to help them stick to the walls and ceilings.

An organic chemist might curdle milk in much the same way you just did as he studies the process of how milk can become glue, cheese, or buttons. He might investigate the curd simply to answer a question. On the other hand, he may want to find a way to improve the process.

You might wonder why the chemist concerned with all this is called an "organic" chemist. At one time it was thought that certain combinations of hydrogen and carbon (called "hydrocarbons") could be made only by living organisms. Therefore the chemist who worked with these compounds was called an "organic" chemist. Today chemists can make hydrocarbons in the laboratory from raw materials that have nothing to do with living organisms. The meaning of "organic chemist" has changed until today it means a chemist who works with compounds that contain carbon. The hydrocarbons and their derivatives form a big family of chemicals. There are probably more than half a million different hydrocarbons and derivatives.

Organic chemists try to improve the natural materials we use. This is how rayon was first made. The chemists who developed it were trying to produce a material to take the place of silk. Today rayon is made from wood pulp and parts of cotton that are treated chemically and formed into fibers.

Nylon was the first completely man-made fiber, in that none of the raw materials was organic in origin. Today some nylon is made from coal, which is, of course, organic. The molecules of nylon usually cannot be found in nature; they are made only by man. This remarkable material is the direct result of the work of organic chemists.

The development of nylon is a good example of how research can often lead, not only to new products, but to whole new industries. It all began when Dr. Wallace H. Carothers started a new program of basic research at the Du Pont company in 1928.

The purpose of the research was, not to work on developing new products or processes, but rather to find new and perhaps fundamental information that could add to the general knowledge of chemistry. One of the things the staff of the new department started to investigate was giant molecules. They wanted to know just how they formed and why. These giant molecules are the important parts of such materials as rubber and cotton. One day one of the chemists pulled on the chemical under test and found that it stretched like taffy when it was warm. After it had cooled, it could be stretched even further and became more plastic and stronger. See illustration on the facing page.

This was very unusual for such material, and more tests were made. It turned out to have remarkable properties that made it ideal as a fiber to be used in weaving cloth. Commercial production of the fiber was begun in 1939. Since that time, of course, hundreds of other uses for nylon have been developed.

To give you an idea of how much a part of your daily life the work of the organic chemist has become, here is a list of the raw materials from which plastics are made:

Air	Petroleum
Coal	Salts
Water	Skim milk
Wood	Acids
Cotton	Natural gas

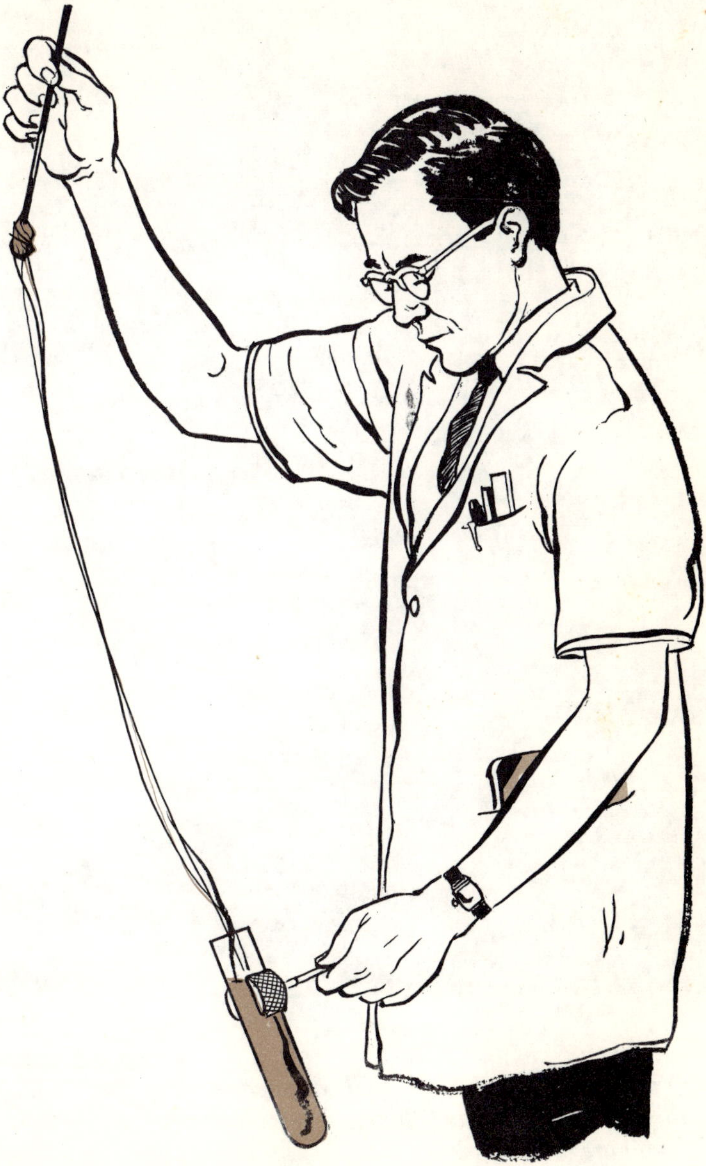

An organic chemist checks the reaction of a chemical experiment. Who can tell, perhaps he will discover a substance as useful and important as nylon. (Photograph courtesy of E. I. du Pont de Nemours & Co.)

And here are some of the most important plastics the organic chemist has made from them:

Synthetic Resins: Many of the common plastics you use in your everyday life are synthetic resins that have been made by the organic chemist. Nylon is probably the most familiar. Bakelite, Plexiglas, Lucite, and the vinyl plastics are also synthetic resins in common use.

The synthetic resins are frequently combined with other materials either as a filler or for other purposes. This greatly extends the number and properties of the final products. In this group of plastics are such everyday items as: radio cabinets, electrical outlets, kitchen table tops, fabrics, raincoats, automobile parts, plastic garden hoses, electrical insulation, adhesives, dyes, medicines, inks, fertilizers, perfumes, insecticides, and so on. These resins are found in the most unlikely places. When a paper tea bag is soaked in hot water, why doesn't it fall apart as ordinary paper would? It's coated with synthetic resin that holds the fibers together.

Cellulose Plastics: Cellulose is the material that makes up the cell walls of plants. The cellulose makes the roots, stems, branches, and other parts of the plant stiff, and this helps support the plant. The plastics that have a cellulose base are very numerous. Celluloid and cellophane, as their names indicate, are cellulose in origin. Cellulose, properly treated by organic chemists, becomes rayon, safety film, pen and pencil barrels, steering wheels, toys, lacquers, eyeglass frames, telephones, combs, etc.

Protein Plastics: Soybean meal is high in protein and is used for making some things, but the best-known protein plastic is the one you know something about: casein.

Next time you need some glue, remember your experience as an organic chemist and head for the refrigerator.

6. What Are the Chances?

Mathematician

If you start a marble at the top of the board as shown on the next page, it will roll down, bumping against the nails, and end in one of the slots at the bottom. Which slot? You might suspect that there is an equal chance that the marble will roll into any one of the eleven slots. In that case, the chance that it will roll into slot 1 is one in eleven. This seems like a very logical conclusion, but it's not correct. The chance a marble will fall into slot 1 is one in 1024!

You'll see why this is as you learn more about how a mathematician works . . . especially in the field of probability.

Before you begin to construct what should prove to be a very interesting mathematical xyzactyu, you should know something about probability. That funny word you just puzzled over is an example of how accustomed you are to expect things are going to happen in a certain way. So seldom do you come across such a jumble of letters in your reading that

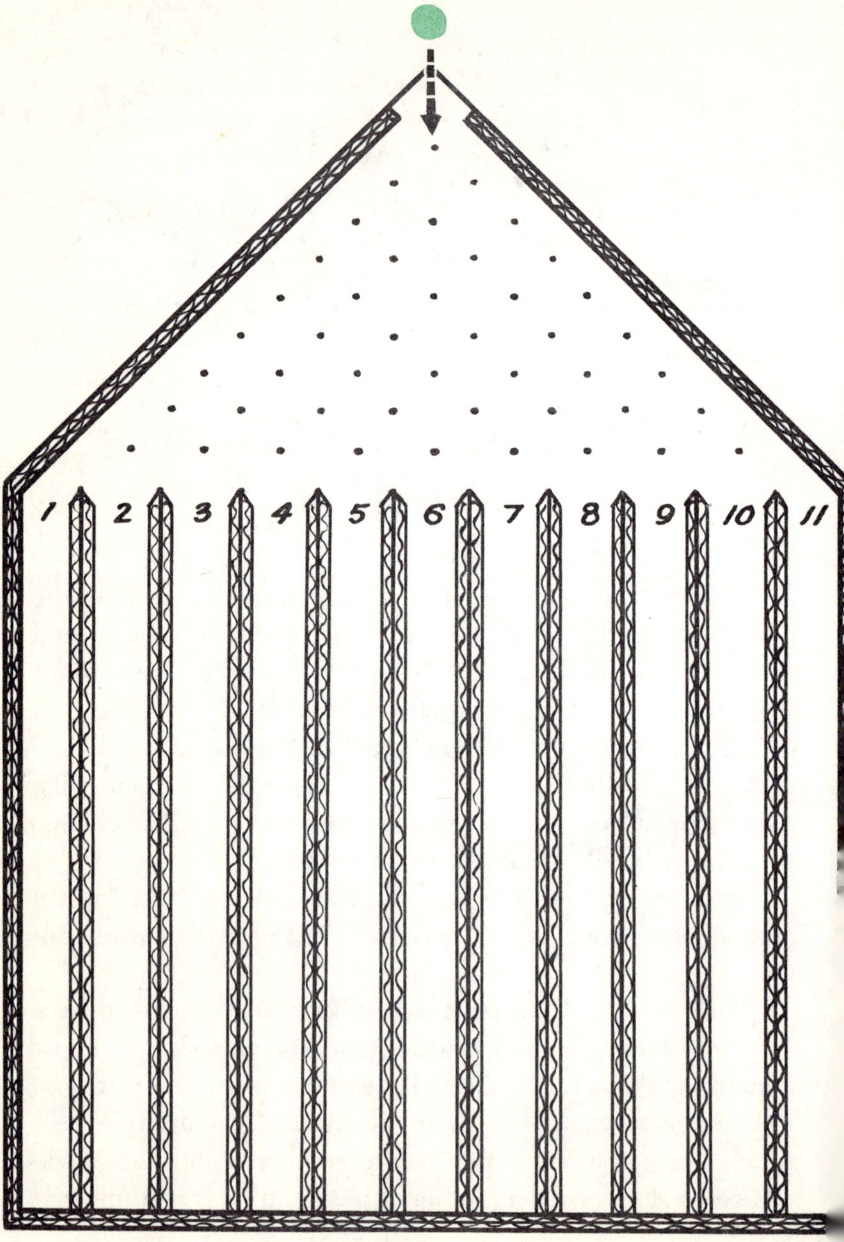
1
2
3
4
5
6
7
8
9
10
11

when you did, you probably stopped to puzzle over it. Before a book is printed, many people have examined it to prevent such a meaningless series of letters from being printed on a page. It is very unlikely that all those people will overlook such an odd group of letters. Therefore you don't expect to see such things in books.

The probability of your having an accident is also remote, even though the newspapers seem to report an ever-increasing number of them. At those intersections where there is more of a chance for an accident to occur, traffic experts have installed stop signs, white lines, caution signs, and perhaps a policeman to warn drivers and pedestrians that special precautions must be taken. All the precautions are taken to reduce the chances of an accident's happening.

You probably don't even realize how often you take into consideration the idea of probability. If someone asks you where he can find a policeman, you could answer: "You might find one at Fourth and Main." Your use of the word "might" indicates that the probability of finding a policeman there is not very good. Here are some other words you use to show various degrees of probability:

certain	uncertain
undoubtedly	mysterious
sure	questionable
without fail	who can tell?
no doubt	depends
of course	incredible
inevitable	inconceivable
unavoidable	impractical
possibly	unthinkable
maybe	unlikely
perhaps	doubtful

With such words as these, you can express very fine differences in predictability.

Now try something that's *un*predictable. Toss a coin. While it's up in the air, call out whether it will land heads or tails up. What are the chances that you will be right? You can be either right or wrong because there are only two possibilities. The chance of your being right is one in two. Mathematicians express the mathematical value of a head's turning up as 1/2.

While it's true you can't predict whether the coin will land heads or tails up, you can be fairly certain that if you toss it a sufficient number of times, half of the total will be heads and half will be tails, because there are only two possibilities. All you have to do is toss the coin into the air often enough, and the totals will get closer and closer to half and half.

If you now toss three coins into the air, the number of possible combinations of heads and tails is increased. There are now eight possible combinations:

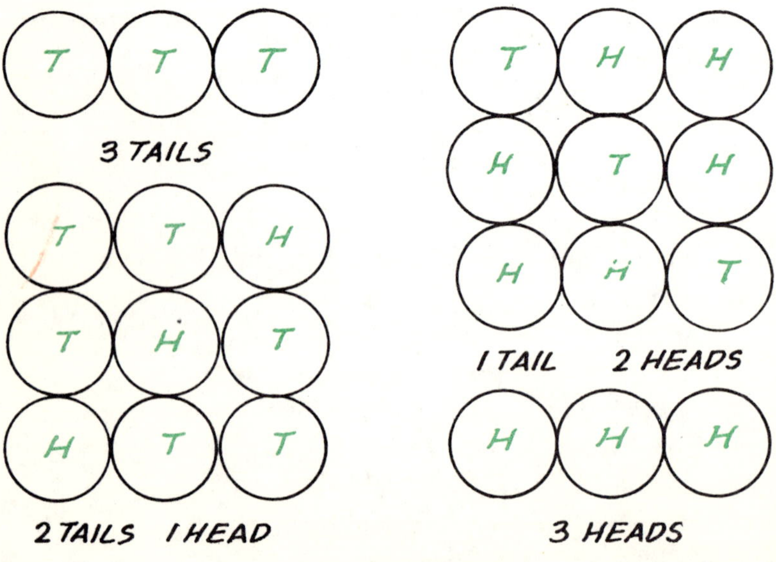

You can see from the illustration that there are three combinations that contain two heads. The chance that the result will be two heads and a tail is 3/8 because there are three possibilities in a total of eight. If you continued to toss the coins long enough, the totals would get closer and closer to these ratios:

tails tails tails	1/8
heads heads heads	1/8
heads tails tails	3/8
tails heads heads	3/8

An interesting extension of this kind of investigation into the mathematics of probability was done by Count Buffon back in the 1700s. You can repeat his experiment by ruling off a piece of paper with a series of parallel lines that are as far apart as the toothpick you are going to use.

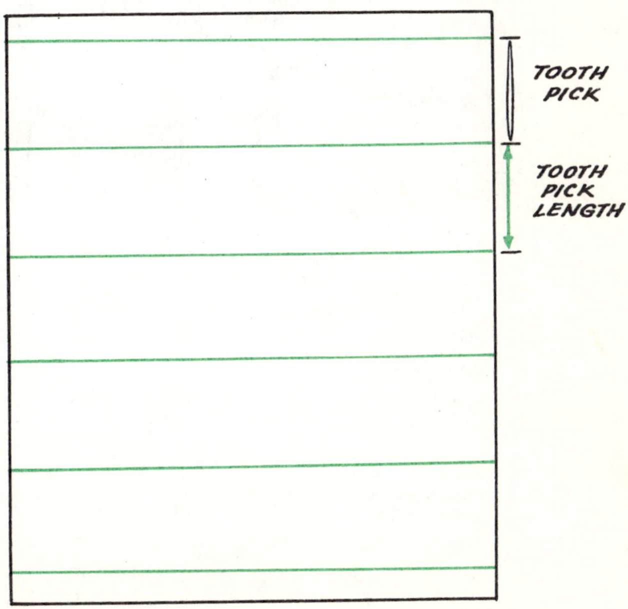

Hold the toothpick over the edge of a chair, or some other convenient height, and let it fall onto the ruled paper as shown:

Record the number of times any part of the toothpick touches any line. Also record the number of times the toothpick does not touch any line. Count Buffon found that if you drop the toothpick enough times, a definite relationship exists between the two possibilities. The chance that the toothpick will touch a line is 2/3.14 or $2/\pi$. You've probably learned that the circumference of a circle is equal to its diameter multiplied by π. You probably also know that the area of a circle is equal to its radius squared multiplied by π. It is strange that this constant so well identified with a circle should appear in what seems to be an unrelated problem.

The greater number of times you drop the toothpick the nearer the relationship of the two possibilities will be to 2/3.14. An Italian mathematician by the name of Lazzerini back in 1901 repeated Count Buffon's experiment. He dropped what corresponds to your toothpick 3408 times. The value he got for π came to 3.1415929. This is only an error of .0000003!

Mathematicians have developed a general formula for this kind of probability. If EL is the number of times that something is *Equally Likely* to happen, and F is the number of times the results could be *Favorable*, then the formula for the Probability of Favorable results occurring could be written: PF=F/EL.

One branch of mathematics that deals with the problems of probability is statistics. The mathematician who works in this field usually begins with a collection of facts which he arranges in some order. This is what your teacher does when she makes a graph to show the results of an examination. She might make up a chart that would look like the one on page 84.

Graphs like that are made by statisticians to quickly organize many other kinds of facts. For example, a very similar sort of chart might result if you collected enough of the following sort of facts and arranged them in a numerical order:

The height of all the boys in school
The height of all the girls in school

NUMBER OF STUDENTS
50
45
40
35
30
25
20
15
10
5
0
50 55 60 65 70 75 80 85 90 95 100
GRADES

The height of all the children in school
The number of bullets that will fall within a certain ring of a target
The velocity of the molecules of a gas
The length of life of men who were 35 years old in 1900

Life insurance companies are particularly interested in graphs based on the life expectancy of various groups of people. Mathematicians, using the information which can be gained from such charts, can calculate how long a person can be expected to live . . . on the average. This information is used as a basis for writing insurance policies.

Karl Friedrich Gauss, a brilliant German mathematician who lived in the late 1700s and early 1800s, worked out the mathematics of such charts and their resultant curves. He formulated a law which is named after him. Whenever facts can be organized according to Gauss's Law, the curve that results will be pretty close to this curve:

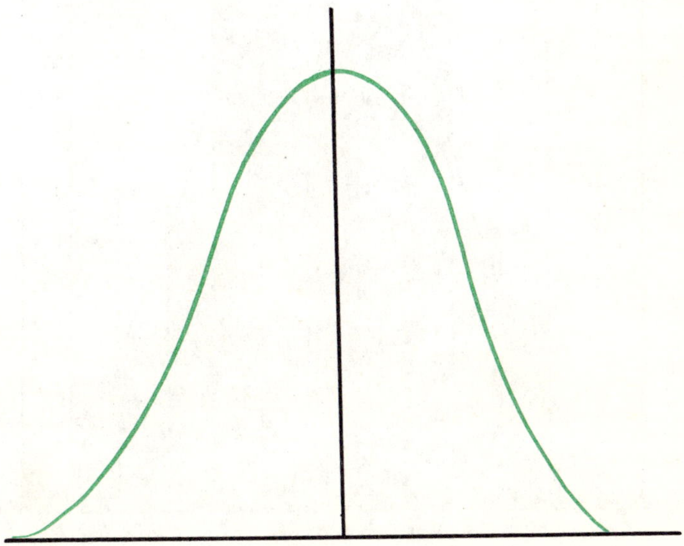

GAUSS'S CURVE

If there were an unlimited number of pupils who took the examination, the chart your teacher made up would follow Gauss's curve. In other words, the above curve is an ideal curve. In practice, the results usually only approach such a smooth curve. The larger the sample however, the smoother the resultant curve becomes and the more nearly it approaches Gauss's curve.

The following figure shows a chart of the lifetime of 290 typical light bulbs. It resembles Gauss's curve.

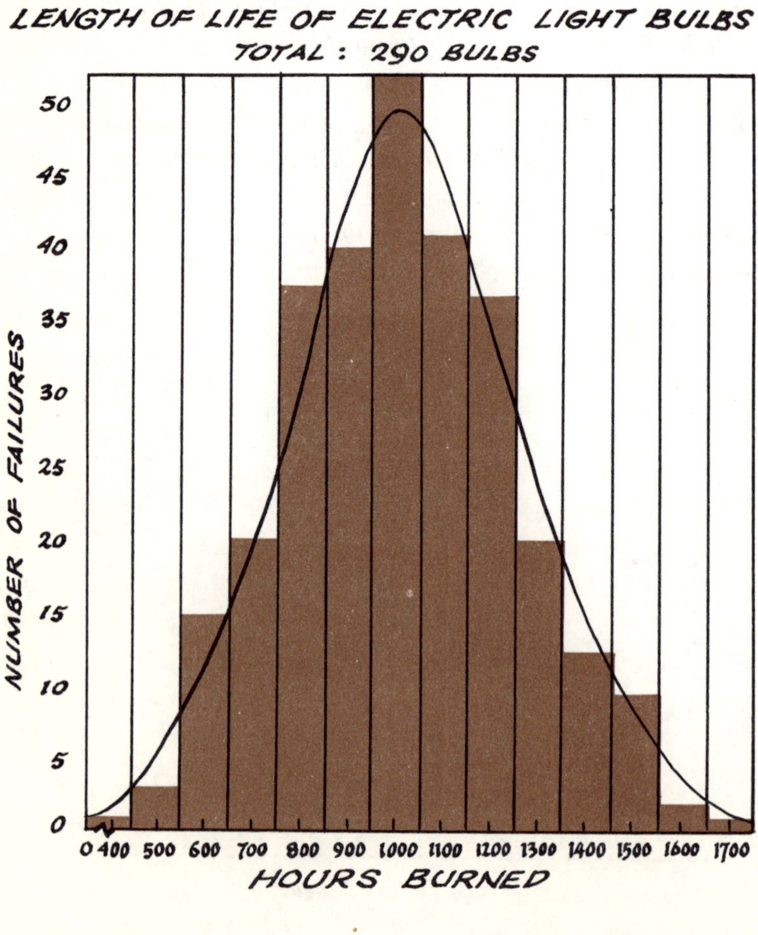

The chances that any individual light bulb will burn 1600 hours is about 2/290 or 1/145. The probability that a specific bulb will burn 900 hours is 40/290 or about 1/7. Note that a large percentage of the bulbs fall into the 800- to 1200-hour groups. To find the probability that a bulb will fall into this section, add together the bulbs included in the section.

HOURS BURNED		NO. OF BULBS
800	–	37
900	–	40
1000	–	52
1100	–	42
1200	–	36
Total		207

$$\frac{207}{290} = \frac{1}{1.428}$$

The answer is 207/290. This reduces to 1/1.4, not far from 1/1, meaning that almost every bulb will burn in the 800- to 1200-hour range. With this kind of probability on which to base their predictions, the company that manufactures the bulbs can declare their bulbs will last 1000 hours on the average.

Gauss's law and his curve will help you understand how the chance of the marble going into slot 1 is only one in 1024.

Mathematically the probability of marbles falling into the channels can be easily understood if you follow one marble down the board. When it hits the first nail, it can go only to the right or to the left. This means the numerical value of the probability that it will go to the left of the first nail is 1/2 (the number of Favorable events possible divided by the number of Equally Likely events).

When the marble hits the nail in the second row, it now has two more possibilities: left or right. The chance that the marble will be on the left side of the nail in the second row is also 1/2. But check the following figure. You'll see the total possible position at the second row of nails is four.

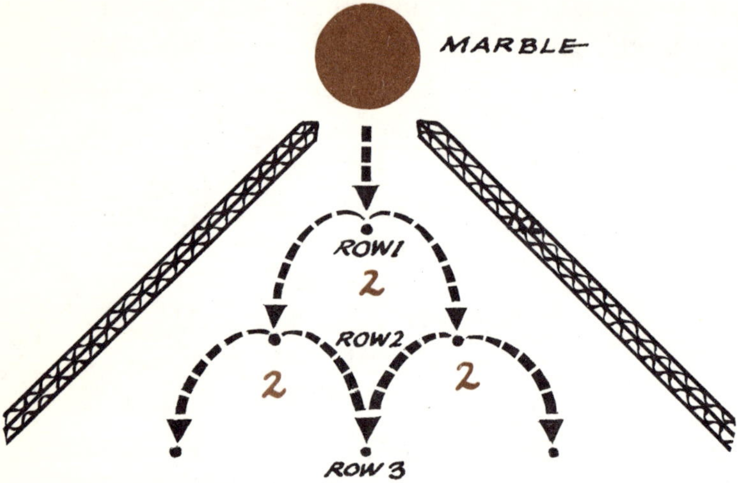

The numerical value for the probability the marble will be to the left of the left nail in row 2 is 1/4. Remember, PF= F/EL. Now the marble bounces to row 3. Here again at the leftmost nail the chances are 1/2 that the marble will bounce to the left. But if you'll follow through the diagram for all the possible bounces at row 3, you'll see that there are eight different combinations.

Note that there are only two possibilities at the end nails, but four possible bounces from the center nail in the third row. The probability that the marble will be to the left of the left nail is 1/8. This same value applies to the probability that the marble will be to the right of the right nail. The probability that the marble will be to the left of the center nail is 2/8. The same value applies when you consider the bounces to the right of the center nail.

When the marble goes onto the fourth row, the number of Equally Likely events doubles. The chances the marble will be to the left of the left nail are now 1/16. The values for the various positions can be arrived at by counting the number of bounces at any one nail.

The black fractions in the following figure on page 90 show what the probability is that the marble will bounce to the left of each nail. The fractions in color indicate the probability the marble will be to the right of the nail. If you add the probabilities in any horizontal row, the total will be one.

The numbers in the slots at the bottom indicate the probability that a marble will roll into that slot. The number is arrived at by adding the probabilities from both of the nails that could bounce a marble into that slot. For example: In slot 4, the nail between slots 3 and 4 could send a marble to the right and into slot 4. The probability of this happening is shown to be 36/1024. The nail between slots 4 and 5 could send a marble to the left and into slot 4. The probability of this happening is 84/1024. The probability that a marble will end in slot 4 is the sum of those two possibilities, or 120/1024.

If you were going to roll 1024 marbles down the board and wanted to make a graph of what to expect, you would arrive at a graph that would look like the one on page 91.

Now that you have calculated the possibilities, you should check the results of your calculations experimentally. The details of the construction of the board are shown on page 92.

The pieces of cardboard can be cut out with a sharp knife. They can also be sawed out with a hand saw if you put several pieces of cardboard together and cut them at once. Pieces of half-inch wood can be cut to three-quarter-inch

$$\frac{1}{2}\cdot\frac{1}{2} = 1$$

$$\frac{1}{4}\cdot\frac{1}{4} \quad \frac{1}{4}\cdot\frac{1}{4} = 1$$

$$\frac{1}{8}\cdot\frac{1}{8} \quad \frac{2}{8}\cdot\frac{2}{8} \quad \frac{1}{8}\cdot\frac{1}{8} = 1$$

$$\frac{1}{16}\cdot\frac{1}{16} \quad \frac{3}{16}\cdot\frac{3}{16} \quad \frac{3}{16}\cdot\frac{3}{16} \quad \frac{1}{16}\cdot\frac{1}{16} = 1$$

$$\frac{1}{32}\cdot\frac{1}{32} \quad \frac{4}{32}\cdot\frac{4}{32} \quad \frac{6}{32}\cdot\frac{6}{32} \quad \frac{4}{32}\cdot\frac{4}{32} \quad \frac{1}{32}\cdot\frac{1}{32} = 1$$

$$\frac{1}{64}\cdot\frac{1}{64} \quad \frac{5}{64}\cdot\frac{5}{64} \quad \frac{10}{64}\cdot\frac{10}{64} \quad \frac{10}{64}\cdot\frac{10}{64} \quad \frac{5}{64}\cdot\frac{5}{64} \quad \frac{1}{64}\cdot\frac{1}{64} = 1$$

$$\frac{1}{128}\cdot\frac{1}{128} \quad \frac{6}{128}\cdot\frac{6}{128} \quad \frac{15}{128}\cdot\frac{15}{128} \quad \frac{20}{128}\cdot\frac{20}{128} \quad \frac{15}{128}\cdot\frac{15}{128} \quad \frac{6}{128}\cdot\frac{6}{128} \quad \frac{1}{128}\cdot\frac{1}{128} = 1$$

$$\frac{1}{256}\cdot\frac{1}{256} \quad \frac{7}{256}\cdot\frac{7}{256} \quad \frac{21}{256}\cdot\frac{21}{256} \quad \frac{35}{256}\cdot\frac{35}{256} \quad \frac{35}{256}\cdot\frac{35}{256} \quad \frac{21}{256}\cdot\frac{21}{256} \quad \frac{7}{256}\cdot\frac{7}{256} \quad \frac{1}{256}\cdot\frac{1}{256} = 1$$

$$\frac{1}{512}\cdot\frac{1}{512} \quad \frac{8}{512}\cdot\frac{8}{512} \quad \frac{28}{512}\cdot\frac{28}{512} \quad \frac{56}{512}\cdot\frac{56}{512} \quad \frac{70}{512}\cdot\frac{70}{512} \quad \frac{56}{512}\cdot\frac{56}{512} \quad \frac{28}{512}\cdot\frac{28}{512} \quad \frac{8}{512}\cdot\frac{8}{512} \quad \frac{1}{512}\cdot\frac{1}{512} = 1$$

$$\frac{1}{1024}\cdot\frac{1}{1024} \quad \frac{9}{1024}\cdot\frac{9}{1024} \quad \frac{36}{1024}\cdot\frac{36}{1024} \quad \frac{84}{1024}\cdot\frac{84}{1024} \quad \frac{126}{1024}\cdot\frac{126}{1024} \quad \frac{126}{1024}\cdot\frac{126}{1024} \quad \frac{84}{1024}\cdot\frac{84}{1024} \quad \frac{36}{1024}\cdot\frac{36}{1024} \quad \frac{9}{1024}\cdot\frac{9}{1024} \quad \frac{1}{1024}\cdot\frac{1}{1024} = 1$$

1	2	3	4	5	6	7	8	9	10	11	
$\frac{1}{1024}$	$\frac{10}{1024}$	$\frac{45}{1024}$	$\frac{120}{1024}$	$\frac{210}{1024}$	$\frac{252}{1024}$	$\frac{210}{1024}$	$\frac{120}{1024}$	$\frac{45}{1024}$	$\frac{10}{1024}$	$\frac{1}{1024}$	= 1

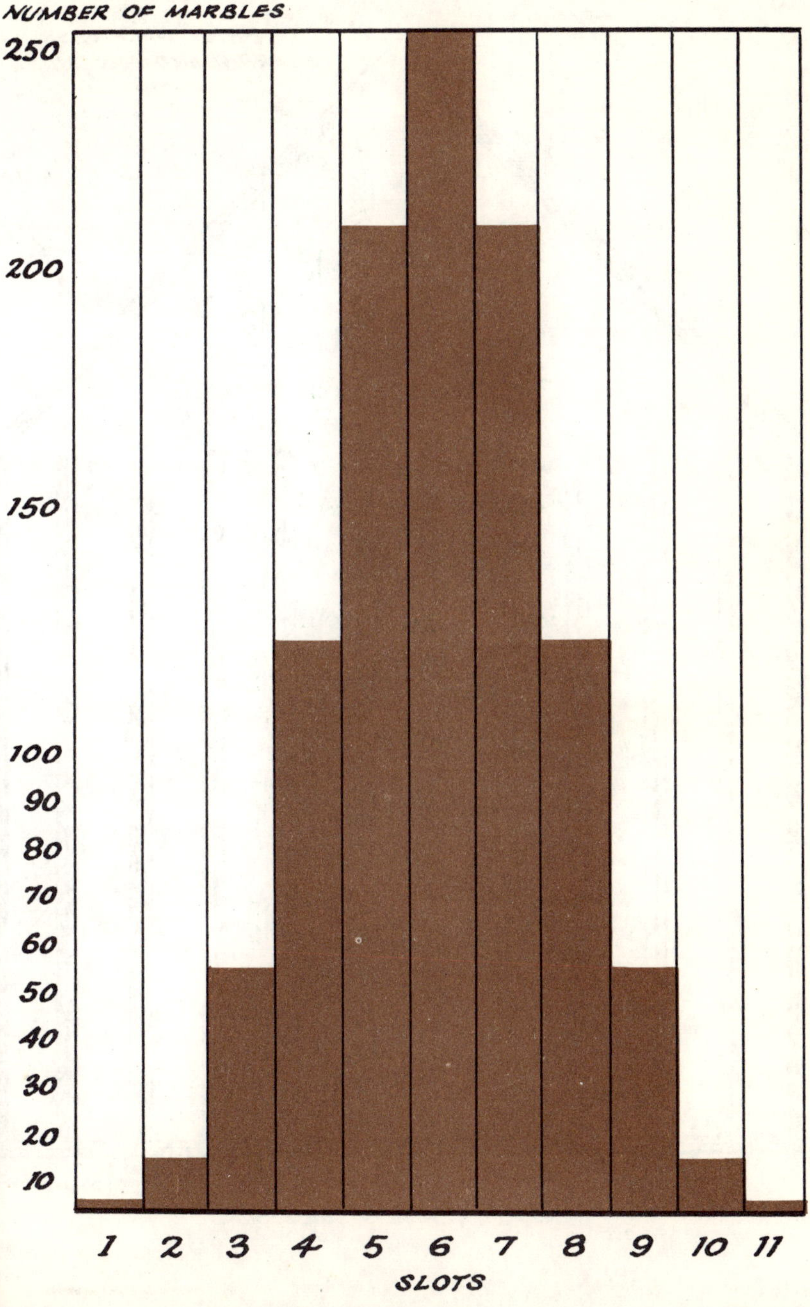
NUMBER OF MARBLES
250
200
150
100
90
80
70
60
50
40
30
20
10
1
2
3
4
5
6
7
8
9
10
11
SLOTS

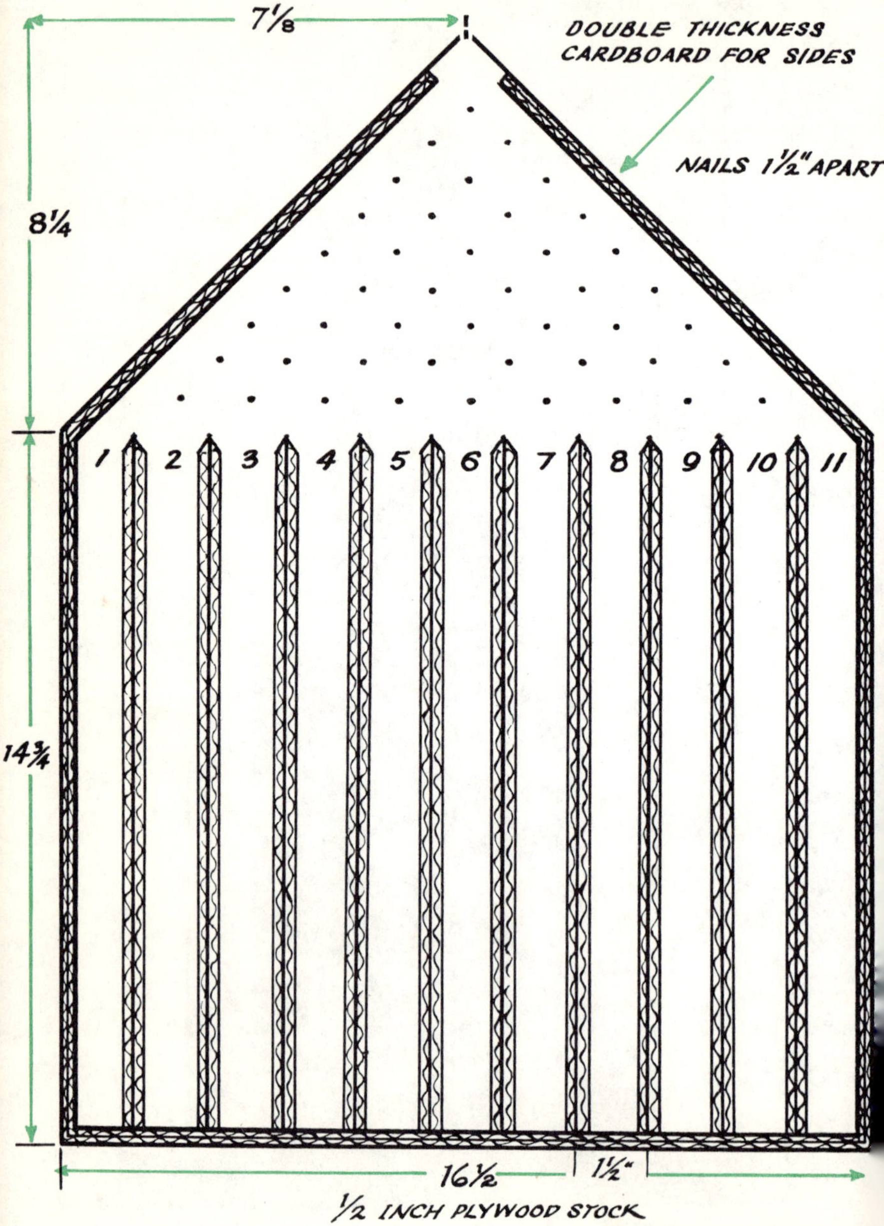
7⅛
DOUBLE THICKNESS
CARDBOARD FOR SIDES
NAILS 1½" APART
8¼
14¾
1
2
3
4
5
6
7
8
9
10
11
16½
1½"
½ INCH PLYWOOD STOCK

widths and the proper lengths and used in place of the corrugated cardboard.

Try to drive the nails into the wood as straight as possible. Also drive them all in to the same depth. Make a wooden stop as shown in the next figure. Drive the nail home gently until the hammer hits the stop.

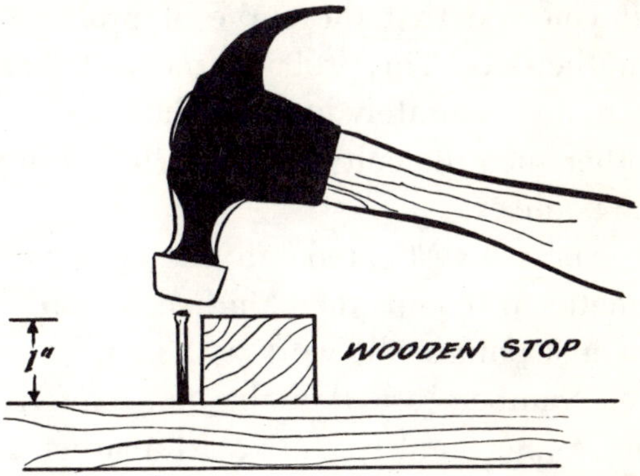

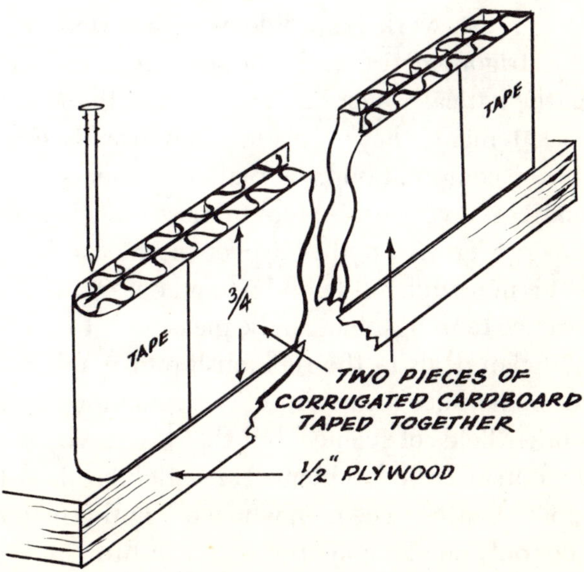

Support the board as indicated about eight inches at the top end and as level as you can. If you have a spirit level, use it to make sure the board does not slant off to one side. A slanted board will affect the way the marbles bounce.

The marbles that should be used are sold in toy and dime stores. They should be made of glass and no larger than five eighths of an inch in diameter and no smaller than a half inch.

What if you find that the curve of probability does not show up in the slots? This probably means the nails are bent, the nails are not accurately located, the board is not smooth, or some other such difficulty is interfering with the randomness of the bounces.

So far you have investigated only one small part of the work a mathematician could do. Mathematicians are usually thought of as scientists who work with numbers, the language of science. Numbers are thought of as referring to volts, beans, guinea pigs, light years. Not all mathematicians work with numbers that refer to something concrete.

When scientists work out problems with arithmetic, algebra, geometry, trigonometry, and calculus, they are often using these mathematical systems as a sort of shorthand for dealing with measurements they've made. They usually can test the results of the computation by actual experiments. In this way, mathematics is a very useful tool to most branches of science.

However, there is another aspect of the mathematician's work that is more difficult to understand because it seldom has any reference to things that can be measured. In this area, the mathematician deals in the relationship of numbers to other numbers. Sometimes the results of these manipulations can be used in other fields of science, but that is not why the mathematician pursued the problem. He is delving into the field of pure mathematics. The men who work in this field consider it to be not only challenging, but also beautiful.

Mathematicians and engineers have designed and built computers like these to solve mathematical problems quickly and accurately. (Photograph courtesy of General Electric.)

Mathematicians are finding more and more demand for their special abilities in solving problems in engineering, finance, economics, research, insurance, computer development, astronomy, and outer-space exploration.

Next time you plan to meet someone at a specified time and place, consider how high the probability is that the event will occur. You see, you may not be a specialist in the mathematics of probability, but you know a lot about it just the same, very likely, most probably, without a doubt, quite possibly, perhaps, of course, for sure!

7. Grocery Farm

Botanist

You've probably eaten sweet potatoes many times, but did you know that you can grow a vine six feet long from one of them? The next time you eat fresh pineapple, remember you can plant the top and it will grow. You can even "train" a plant to do "tricks"!

Why not become a junior botanist and grow some interesting plants? You'll be surprised how decorative ordinary vegetables can be. To start your own botanical garden, you'll have to have seeds, but not necessarily the seeds that come in little paper envelopes. In fact, it's more fun to get the seeds in their natural "packages." Buy a few cents' worth of peas in a pod and you're ready to start a garden on your window sill.

A convenient container to grow them in is a waxed paper cup as shown on page 98.

Punch a few holes in the bottom of the cup, fill with dirt, put it in a saucer, and you'll have an ideal place to begin to

grow seeds. Later on, you can transfer the plant to ordinary pots or larger containers. Most seeds will begin to grow faster and better if kept in a dark place until they begin to sprout. Then move them to a place where they will get plenty of sunlight. Water your plants now and then, but not too often. If the dirt is moist, probably there is enough water. Watering seeds too much sometimes will make them decay rather than grow. If the peas you get are dry, soak them for twenty-four hours before planting.

Once the plant has started to grow, you can add plant food or "fertilizer" from time to time. You can get fertilizer in places where they sell potted plants. Be sure your plants don't get too cold. If you keep them inside, it's a good idea to clean the dust off the leaves every once in a while.

Here is a list of seeds you can find in a grocery store, with hints for growing them: pea, bean (kidney; snap, green or lima), corn, and lentils grow very easily. Of course, you can't grow the canned variety because they've been cooked.

You can get many seeds from just one raspberry, strawberry or blackberry because each fruit is actually a carrier for many small seeds. Sometimes you can feel them when you eat the fruit. If many small plants grow, wait until they are well established and then weed out the weaker ones, leaving only a few stronger ones to continue to grow. Blueberries, also, can be planted. Plant several blueberries in one cup until they can get started.

Pecans and walnuts in their shells will grow into large trees if you give them time enough. In order to get them to grow, soak them in water for several weeks, and then leave them in a dark place to sprout. Once they begin to grow, you can transfer them to pots. Eventually you'll have to plant them outside because they'll be so big. Most seeds with hard coverings, like nuts, will take a long time to grow. Peanuts from the grocery store have probably been roasted and will not grow.

Orange, lemon, lime, grapefruit, cherry, apple, avocado, watermelon and other melons have seeds you've seen many times. See illustration on page 100. Notice we don't eat the seeds of these "fruits," just the part around the seeds. In the case of cucumbers and tomatoes, however, we eat seeds and all, although not primarily for the food value in the seeds. In this group are trees, upright plants, climbing plants, and vines. All are very attractive to have around the house.

What do you suppose will come up if you plant some birdseed? You can have a lot of fun identifying the plants that you grow. You'll need some help in finding out what they are. Check your library for books on how to identify plants.

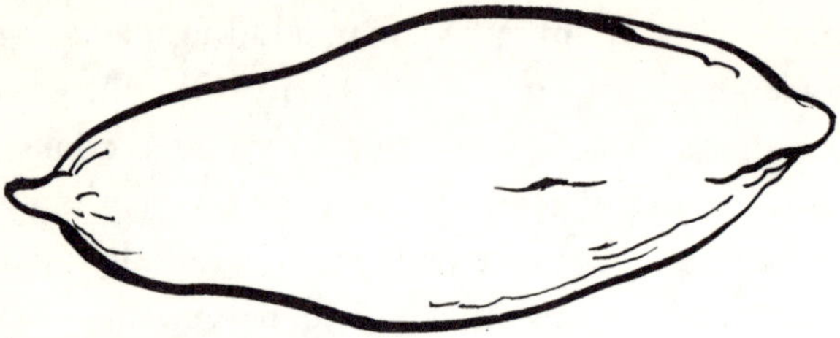

SWEET POTATO

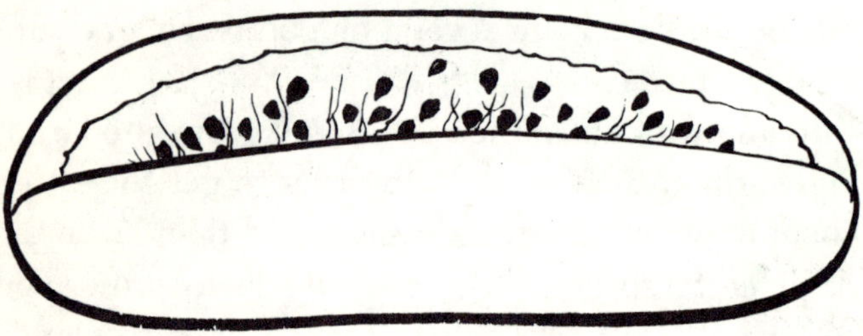

WATERMELON

LEMON

The sweet potato vine is an exceptionally attractive plant. Try to find sweet potatoes that have already started to sprout if you can. In any case, get four or five and put them into jars so that one end is in water. Be sure to add water to the jar from time to time. Keep them in a dark closet until they sprout (which may take several weeks), then move them into the sun. The sweet potato vine can grow to be as much as six feet long.

Carrot, beet, radish, parsnip, turnip all can be grown from seeds, but your grocery store probably doesn't sell the seeds. You can grow the top of any one of these vegetables by cutting it off as shown here:

Cut off the old leaves and place the top between pieces of modeling clay to hold it upright. Keep water in the bowl, and, before long, new leaves will grow out from the top. These plants have large roots with food stored in them. We eat them to get the food stored in the roots. The leaves will grow until the food in the top of the root has been used up.

Onions of various kinds can also be grown very easily. You can sometimes find them growing in the grocery store. "Plant" them in the same way you did the carrot. The new onion plant uses the food stored in the group of special leaves you call an onion. A top which is especially interesting is the pineapple. Cut off as you did with the carrot, put it in water, and watch it grow.

Be sure to label each cup or pot with the name of the plant, write the date that you planted it and any other information which you think might be useful to you later on. Another system is to number each pot and keep the record on another sheet of paper according to numbers. If you don't know what the seed is, describe it as well as you can before you plant it. You can use your description of the seed later to help you identify the plant when you look it up in a book in the library. This is exactly what the professional botanist does when he tries to identify plants.

Botanists have carefully studied every one of the plants that you grow. Sometimes botanists make a special study of the plants that grow in their area and are called upon for advice concerning these plants. Even the weeds along the road have been carefully examined and classified.

Some botanists travel thousands of miles to the jungle or arctic to find out about the plants that grow there. These botanists want to know how plants grow "in the wild." Sometimes they discover a "new" plant and can have the honor of naming it. So far, botanists have classified and given names to about 330,000 types of plants!

A botanist can work in many subdivisions of the field of botany. Here are some of them: studying plants and their uses, classifying plants, investigating the form and structure of plants—including the cells—mapping the distribution of plants on the surface of the earth, examining the

This botanist is raising organisms that cause plant diseases. She is conducting a series of experiments to find out how these organisms react to different chemicals. She is looking for a chemical that can control the disease. (Photograph courtesy of E. I. du Pont de Nemours & Co.)

chemical make-up of plants, developing methods of controlling plant diseases, finding evidence of plants that are now extinct, and making a special study of only one section of the plant world—such as fungi or grasses.

The study of plants is extremely important to all of us because all of our food comes from plants. We either eat the plant parts directly, as in the plants you've been growing, or indirectly, because we eat the animals who eat the plants for food. For this reason plants have been called the food factories of the world. Plants can "store up" energy from the sun for later use. Ordinarily this process goes on inside the plant cells. However, an American botanist, Daniel Arnon, and his fellow scientists, in 1954, were able to make a plant chemical do this—outside the plant itself. This means that for the first time there is a possibility that botanists may be able to analyze the important steps by which a plant makes food. This could possibly prepare the way for the development of more efficient use of the plant's food-making process.

You won't be investigating such complicated things as you grow your grocery-store garden, but you may want to continue your study of plants. As you progress, you can start a small experimental garden, visit and study in a municipal botanical garden, and "explore" in the "wilds" of wherever you live and wherever you go. Fortunately for us, plants grow almost everywhere.

Most people think plants are living things without much movement or sense. However, plants have both, and you can prove it with an interesting experiment. At the same time you will see how plants respond to light. During this experiment you can make a plant "run around in circles"!

Use one of the plants that grow easily like bean or pea or, better yet, use a potato that has already sprouted. After the plant is established and growing well, put it inside a shoe box on top of another box such as a milk carton. Cut a hole in the center of one side of the shoe box. You don't necessarily have to use a shoe box or milk carton—any combination of boxes that will put the plant into the position shown here will do:

Put the top on the box and place it where a good supply of light will fall on the hole. Open the box every day and record the growth of the plant toward the light. Scientists aren't sure of the exact mechanism whereby a plant responds to the light in this way, but it most certainly does somehow "sense" the light and move toward it.

Before the plant has grown far enough to come through the hole, turn the box, or plant as indicated here:

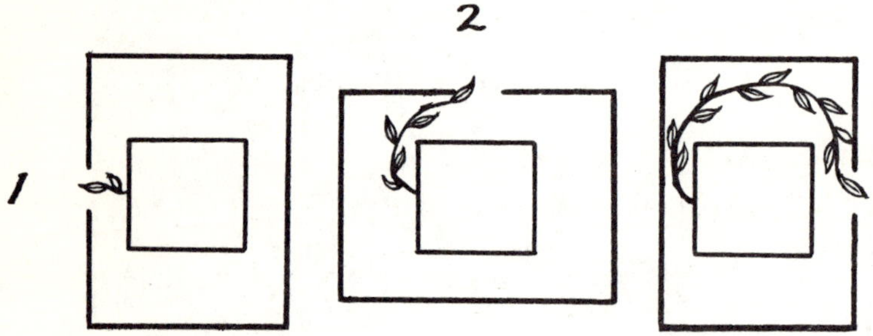

Just as you suspect, the plant will now grow toward the hole again. If you have a hearty plant that grows well, you can change the position of the hole so that the plant will circle around the shoe box.

By putting together the right combination of boxes and holes, you can make the plant grow into all sorts of weird shapes. When you have the plant in the design you like, remove or cut away the boxes and everybody will wonder how in the world you trained the plant to do such strange "tricks."

8. The World through a Drop of Water

Microscopist

Have you ever explored the fascinating world that's visible under the microscope? If you haven't begun this adventure because you haven't got a microscope, why not build one? The most important part is a drop of water. The rest of the material comes from your house and the hardware store!

You see, a drop of water can magnify things because it is shaped like a lens. How well it magnifies is easy to see with the help of a piece of waxed paper.

With a medicine dropper, put a drop of water in the center of a small piece of waxed paper. Place this on top of a newspaper, magazine, or book, and look at the print through the drop of water. You will see the print enlarged. Pick up the waxed paper and hold it a short distance above the print and the print will appear even larger.

Using a drop of water like the one on the waxed paper, you can make a microscope that will magnify objects as much as 100 times and see such things as the dots that make up a picture in a magazine, the fibers that are twisted together to make a thread, the hairs on a fly's legs, and hundreds of other interesting things.

A scientist who uses a microscope could be called a microscopist. But if he studies insects, he usually considers himself to be an entomologist; if he examines plants, a botanist; if he investigates mineral specimens, a geologist; if he investigates crystals, a physical chemist. So you see, as you investigate things with your water-drop microscope you will be many kinds of scientist in the fields of physics, chemistry, and biology.

Making the microscope is relatively simple. Here is a list of the materials you will need.

Piece ¾-inch wood 5¾ x 3 inches
Piece ¾-inch wood 5¼ x 3 inches
4 pieces ¾-inch wood 1 inch square
1½-inch pipe coupling
1½-inch pipe nipple 3 inches long
1½-inch floor flange
"Keyless" (no switch) electric light socket
7½-watt frosted bulb
⅛-inch pipe nipple
⅛-inch lock nut
6-foot length lamp cord
Male plug
2½ x 5½-inch piece sheet metal
Medicine dropper
Small glass of distilled water
4-inch square piece of stiff plastic
Miscellaneous nails and screws

Assembly of the wooden pieces and the pieces of pipe is shown on page 110.

The pipe should first be cleaned with soap and water to remove the oil and dirt. Then the nipple, or straight piece, is screwed into the floor flange. The coupling is then screwed on the end of the nipple. All pieces should turn easily. A bit of vaseline or oil on the threads will help keep this action smooth. The floor flange, nipple, and coupling are then placed directly over the light bulb.

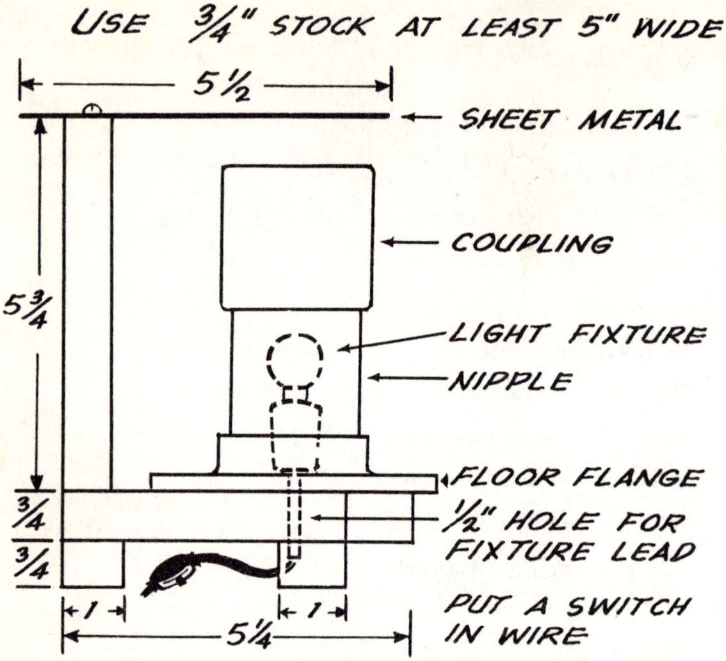

SIDE VIEW

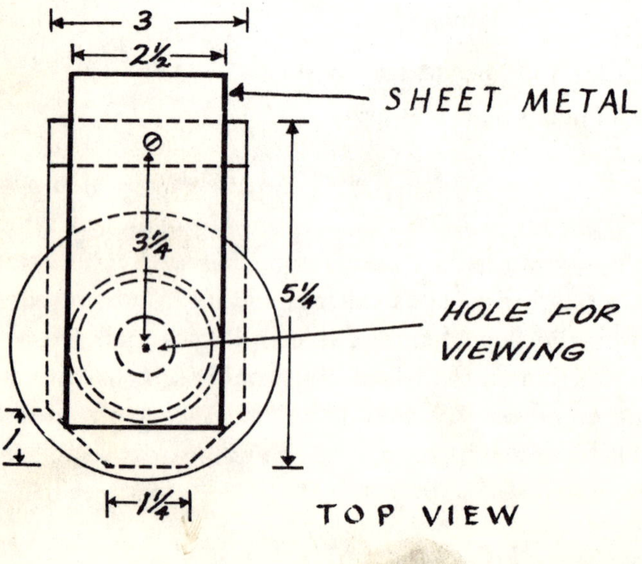

TOP VIEW

If you have had no experience in assembling a light socket and a male plug, better get some help from someone who has. Putting the proper electrical equipment on each end of the cord is not difficult but must be done properly.

The metal piece that supports the magnifying water drop needs some special attention:

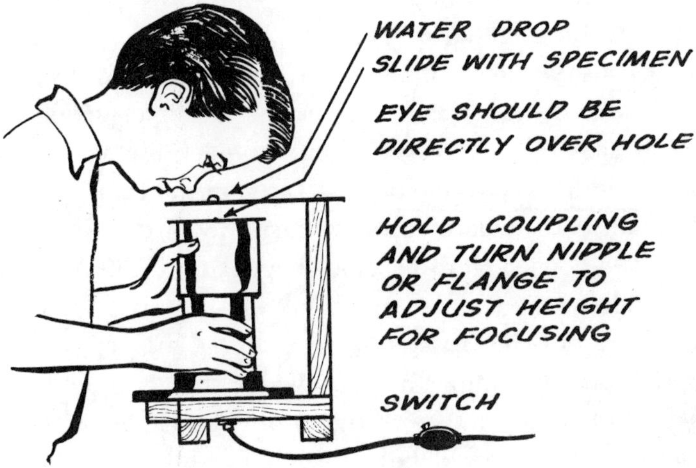

Make it out of the thickest metal that you can conveniently work with. A piece cut from a tin can will do. Drill a one-sixteenth-inch hole as indicated opposite. File both surfaces of the metal to remove any burrs and to get a smooth round hole. Coat the end of the sheet metal with oil, lard, butter, petroleum jelly, or similar material and then wipe it off slightly, leaving a thin film on both sides. With a toothpick, clean out the hole. Now, mount the metal piece with one screw. Do not tighten the screw all the way. You should be able to move the metal arm with a slight pressure.

The piece of plastic must be large enough to cover the end of the pipe coupling. The specimens you are going to examine are placed on this piece of plastic. The plastic can come from a tool or cheese box, or a small piece of glass can be used if no plastic is available. The important thing is to have a transparent covering for the pipe on which you can place the specimens to be examined.

It is best to use distilled water for your lenses. You can get distilled water in the grocery store. It's used in some steam irons because it has no minerals or other materials dissolved in it. This is why it is also best to use in your microscope. The material that can be dissolved in ordinary tap water may interfere with the passage of light through the drop.

Can you guess where you have a supply of distilled water around the house? Here's a clue: it's not in a bottle and it's frozen. That might make you think of the refrigerator and you'd be right. The frosty ice-like material around the coils near the freezer compartment is water that has condensed there out of the air in the refrigerator. This, in effect, is distilled water and should be pure enough for your purpose. Scrape some of the frost into a clean glass and let it melt. It will make excellent lenses.

With the medicine dropper, place a drop of distilled water over the hole in the end of the metal. The reason for putting petroleum jelly, butter, or oil on the sheet metal is to keep the water in drop form. The water will not wet the metal as long as the oily material coats it.

Put an ordinary pin on the plastic and turn on the light. Move the pin until the head is under the water drop. Now bring your eye as close as possible to the drop of water. Hold the pipe coupling with one hand and do not let it turn as you turn the pipe. If this does not bring the pin into focus, hold

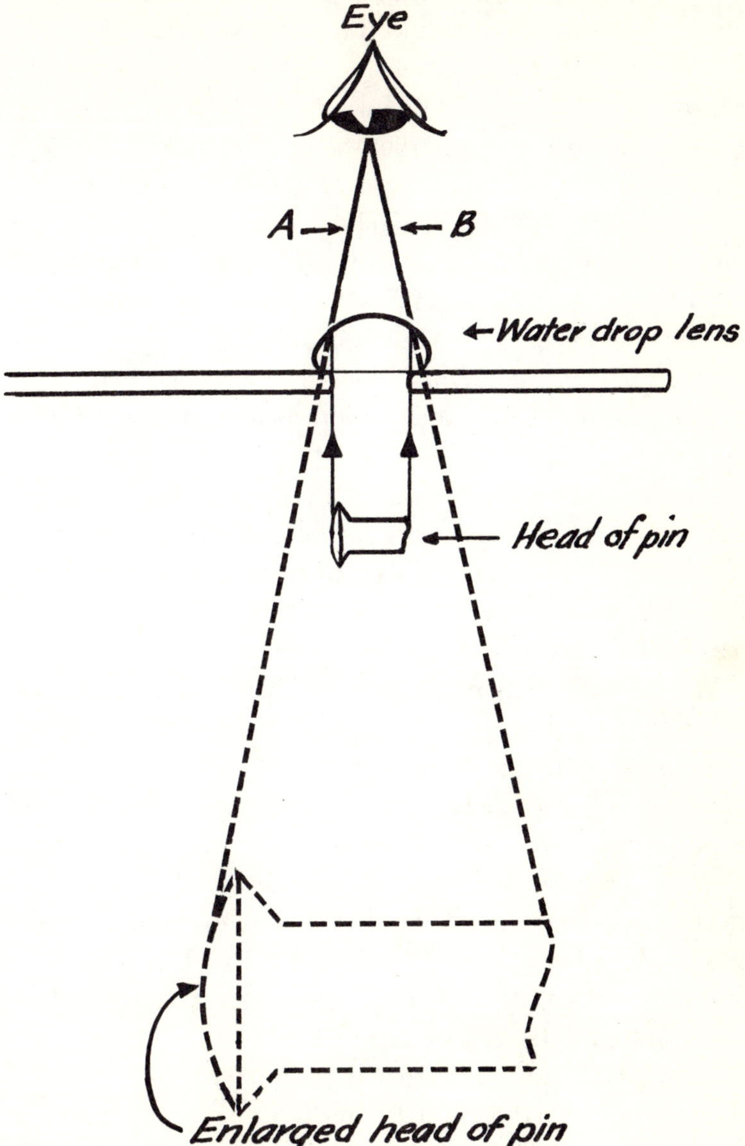
Eye
A
B
Water drop lens
Head of pin
Enlarged head of pin

the nipple as you turn the pipe flange. By this method you can bring the pinhead closer or take it farther away from the water drop. Adjust it until you have the image of the pinhead as clear as possible. You will be amazed at how much you can see with the water-drop microscope when it is properly focused.

If you cannot get the image to be sharp and clear and you're sure you have focused as well as possible, try changing lenses. Gently wipe away the water drop. Carefully put a new drop over the hole, trying to make it a different size from the previous drops. You should have little trouble once you determine the size of the drop and the approximate distance for the image to be in focus.

This is a "microscope" because it is a device for looking at small objects. It does not work in quite the same way as the microscope which scientists use. Most of the time they use what is known as a compound microscope, which means that it has more than one magnifier. Yours works on the same principle as a magnifying glass or hand lens. So that you will be able to get maximum magnification with your lenses, you should understand how they work.

By studying the illustration on page 113, you can see that light reflected from the head of the pin goes through the drop of water where it is bent in such a way that it enters the eye on the line marked A. The eye automatically assumes that if the light from the head of the pin enters along a line such as A, the head of the pin must lie somewhere along a straight line projected from A. This is indicated by the dotted line that runs from A to the imaginary head of the pin. The same thing happens with the light reflected from the shaft of the pin. The eye assumes that the shaft must lie somewhere along the extension of the line B. This is how you

To take this picture the photographer used what amounted to a low-power microscope combined with a camera. Note how hairy the fly's body is and the pads of sticky material on his feet. These allow him to walk upside down even on glass. (Don Ollis–Black Star.)

see a magnified image of the pin through the drop of water.

A large drop of water with little curvature will not magnify as much as a small drop of water with a greater curvature. The light is bent at a greater angle by the smaller drop.

Now that you have built your microscope, what are you going to look at? Well, you can look at practically anything. Remove a hair from your head, or someone else's head if you prefer, and place it on the plastic of the microscope. Adjust it until it's in focus, and you will be amazed to see that the hair looks like a huge rope. If you have managed to pull out the hair by the roots, examine the roots too.

Examine some mold plants with the microscope. (See Chapter Three.) You will find that they look like little balls on the end of a stalk. If they happen to be ripe, you may actually see the tiny spores from which new mold plants grow.

Tear a small section of a color photograph from a magazine and place it under the microscope. You will be surprised to see that the picture is actually made up of thousands of tiny dots of color. By putting together the various dots, different colors are produced.

Place parts of a fly under the microscope and examine them. You will be able to see the hairs on the fly's legs and even the pads on its feet that help it walk upside down on the ceiling. Be sure to examine the wing also.

As you do the experiments in this book, look at the materials you are working with: salt, yeast, pencil point, ants, seeds, etc.

For other suggestions concerning what to look at, go to the library. There you'll find whole books about microscopes and how to use them. Many of the things that are suggested for study with a regular compound microscope you should be able to see with your microscope. Be sure to replace the drops of water from time to time. Remember, your lenses can evaporate!

9. The Rolling Jar That Works

Physicist

When you roll a glass jar down an incline onto a rug, let's say it rolls fifty-two inches and stops. Now, if you raise the incline and move the jar down it until its distance above the floor is still three inches, the incline is steeper. Will the jar roll beyond or short of fifty-two inches? See illustration on next page.

Physicists have already worked out problems similar to this and have devised a formula that should tell you what the answer is going to be before you try it. Here is what the physicists have found: the length of the roll of the jar depends upon how much energy it has at the moment it leaves the incline and the friction it encounters during its roll. If you assume that the friction of the air and the rug remain constant

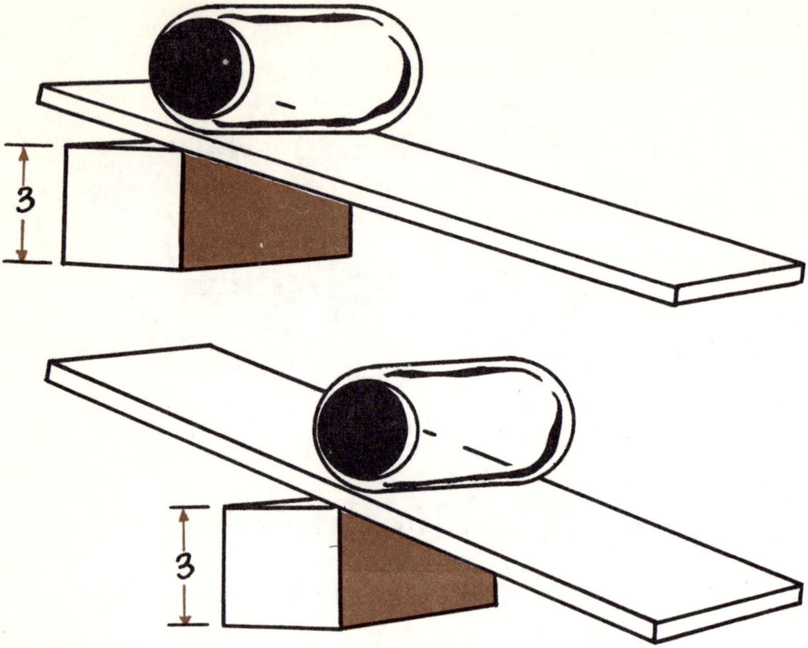

WHICH JAR WILL ROLL FURTHER ?

for every roll, then the determining factor will be the energy of the jar at the end of the incline.

Physicists measure energy by the amount of work it can do. The work in this case is done in overcoming friction as the jar rolls on the rug. The energy to do this work comes from the energy the jar gets rolling down the incline. The jar gets this energy from the pull of gravity. The force of gravity is pulling the jar down just as though it were falling vertically downward from its starting point on the incline to the rug. The incline simply makes this falling take place over a longer period of time.

How much work can the jar do? The physicists measure work in a special way and give it special units. When a force acts through a certain distance, the work done is the force multiplied by that distance. In this case the force is the weight of the jar, and the distance is the height. The units are foot pounds. The formula is E = HW.

The H in the formula is the distance the jar is above the rug. For your purposes, assume the distance is three inches. This is a quarter of a foot. Notice that the distance at the beginning of each roll is the same. As far as the height (the H part of the formula) is concerned, the length of each roll will be the same.

The other part of the formula is the weight. If the jar weighs two pounds, the energy it will have at the bottom of the incline will be ¼ foot x 2 pounds, or half-foot pound. This is the same as the work required to lift the jar this distance. But you haven't weighed the jar and you don't have to. Its weight remains constant during all the trials. As far as the weight is concerned, the length of each roll will be the same. Therefore, with H and W both remaining the same and the friction during the roll remaining the same, the length of the roll should remain the same! This may seem hard to believe, and perhaps you'd better prove it experimentally.

The jar should have a wide mouth and a metal top. A jar for instant coffee or mayonnaise is ideal. The larger size is easier to work with. Run a piece of tape inside the jar as straight as you can.

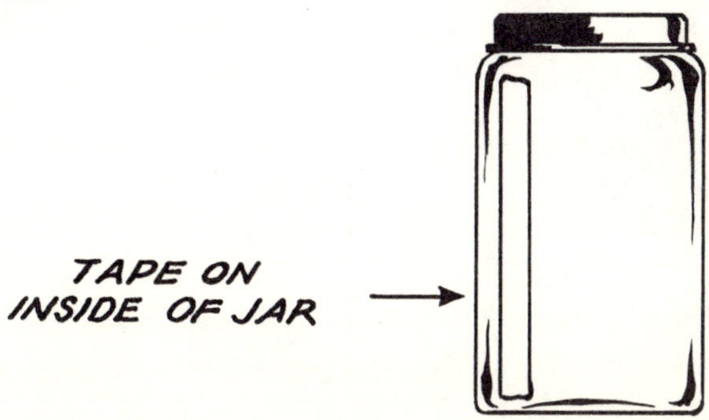

The incline can be made of any stiff material. Cardboard, plywood, masonite, a slat from an apple crate, or any material that's smooth and long enough to make a gentle incline is all that's necessary. Mark the position of the lower edge, so that, in moving the board up or down, it can be kept on this mark.

The easiest way to get the jar three inches above the rug is to mark off three parallel lines on the board before putting it into position.

Make the distance between the parallel lines at least equal to the diameter of the bottle you're using. Use a box or a pile of books to support the board so that the distance from the rug to the topmost line is three inches.

Now, put the jar on the board and line up the tape with the first line by looking through the glass jar. Make a mark on the edge of the jar top directly above the line on the board. See illustration on page 122.

Hold the jar lightly with the fingers as shown on page 123. Then let go.

Allow the jar to roll until it stops. Put a piece of light cardboard or heavy paper on the floor next to the jar and anchor it with weights of some kind so that it can't move. Other bottles or books will do. Make a mark opposite the bottom of the jar on the cardboard. This is the distance the bottle rolled. Measure the distance from the bottom of the incline to this mark. Let's say it's fifty-two inches.

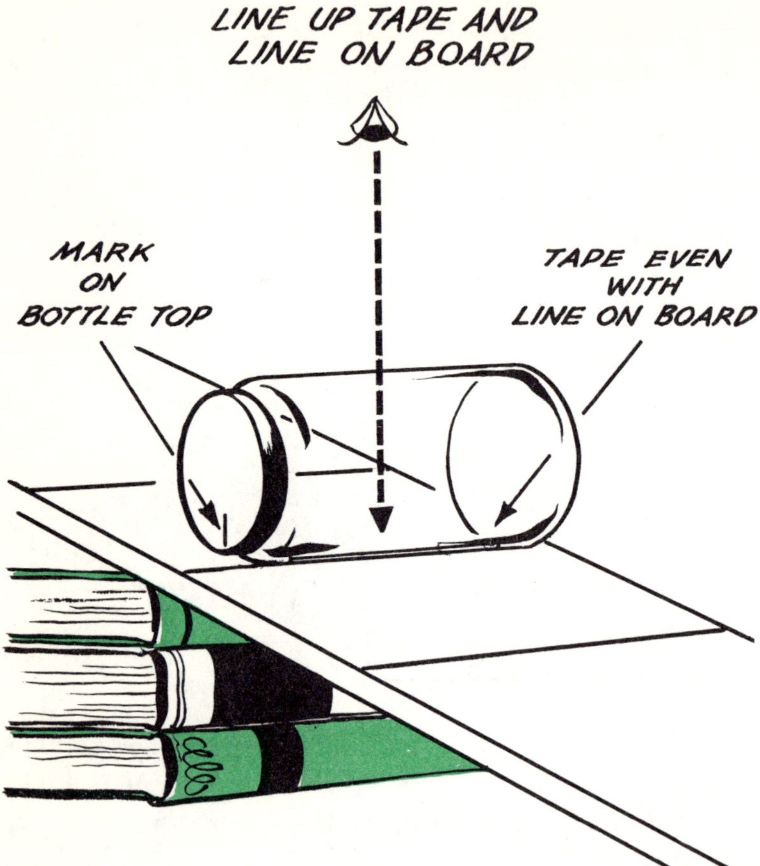

Repeat rolling the jar down the incline, using the same system of lining up the tape with the line on the board and the mark on the jar top with the line at the edge of the board. Record where the jar stops each time. Be sure to record on the cardboard what the height of the incline is and which line you're beginning from.

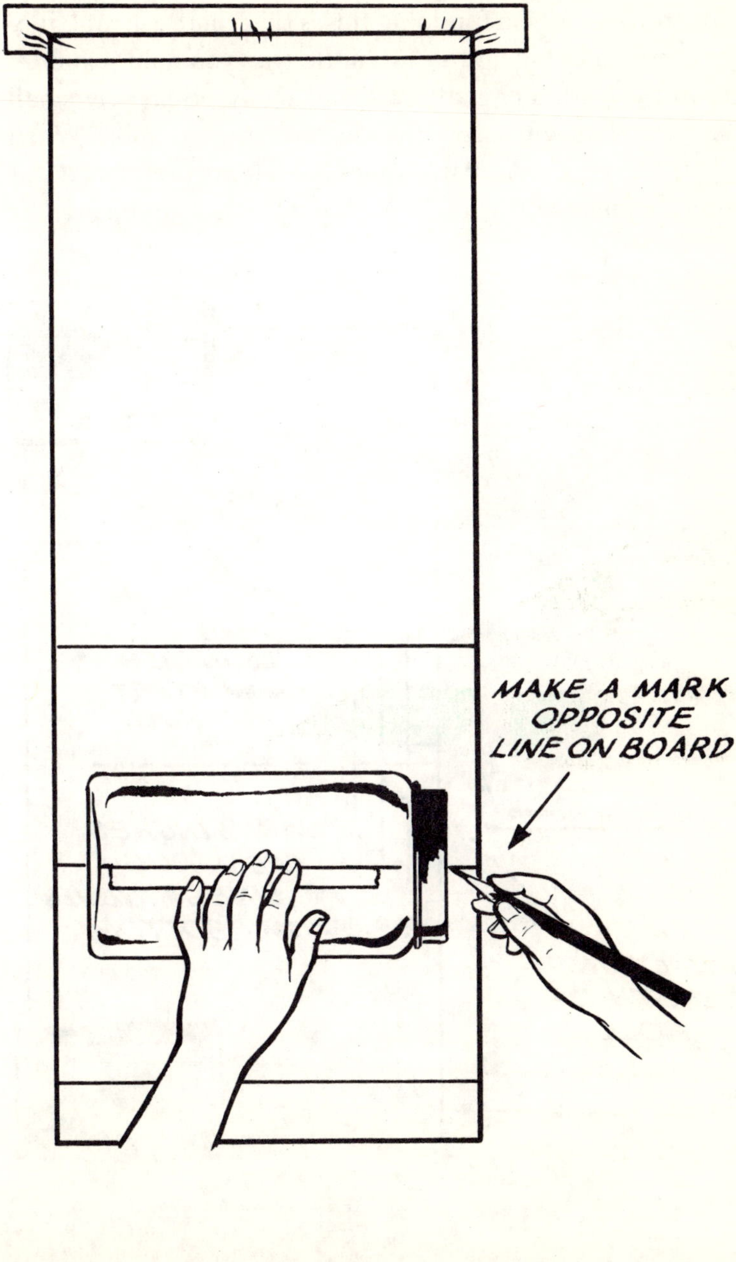
MAKE A MARK
OPPOSITE
LINE ON BOARD

Continue to roll the jar down the incline until you have a good grouping of marks around some kind of center on the cardboard. The reason for doing this is that there are probably slight errors in lining up the jar, in the way you hold it and let it go, in the path along the rug, and many other very small differences that will affect the distance the jar rolls. When you roll the jar down many times, you tend to average out these differences and get a more accurate measurement.

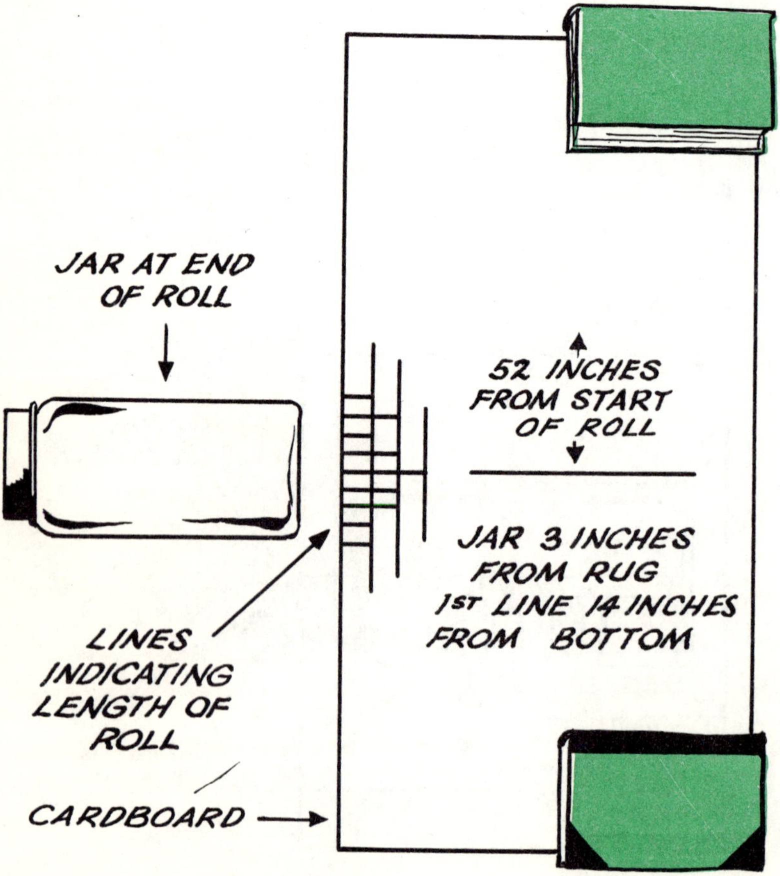

Note in the above figure that the length of the mark indicates how many times the jar stopped at that mark. If the mark goes only to the first line, it means the jar stopped there once. If the mark goes to the second line, it means the jar stopped there twice. Consider the jar to stop at the nearest quarter inch mark and you'll be surprised at what will develop if you roll the jar down the incline enough times.

If you roll it down thirty or more times, you will see a good example of the curve of probability developing. You could even lay the results out in graph form and you might get something like this:

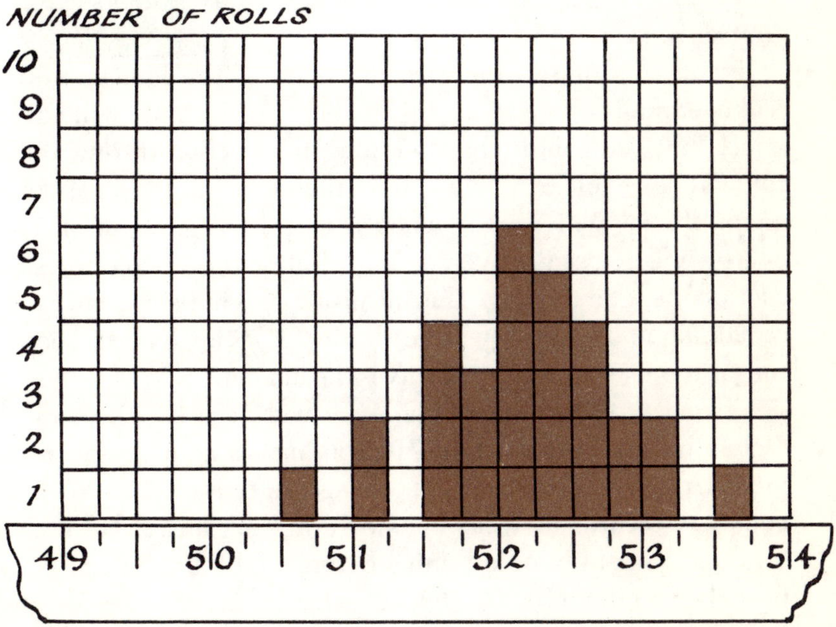

Be sure to read the chapter on the mathematician, because there you'll find an explanation for why the marks fall into this pattern.

By the time you've rolled the jar down thirty times, you

will be able to predict with a good degree of accuracy how far the jar will roll.

Now, change the position of the incline, keeping the front edge in the same place. Raise the rear until the second mark is now three inches from the rug. Repeat the rolling procedure. Mark the new distances on a separate piece of cardboard. Be sure to record the distance from the bottom of the incline. Will the second series of rolls give you the same distance? Remember the formula: E = HW. How about the length of the roll from the third line when the incline will be steeper? Will it agree with the results of the rolls from the first two lines? You'll have to try it and see for yourself.

If you continue to increase the slope of the incline, you will eventually come to an angle where the formula will still apply, but the conditions governing the friction during the roll will be changed. When you set the incline at too steep an angle, the jar will bounce when it hits the rug. Some of its energy will be used in overcoming additional frictional forces as it bounces on the rug. This means there will be less energy remaining and the roll will be shorter. For this reason, keep the jar rolling smoothly as it leaves the incline.

As you do these experiments, you will be using the same techniques a physicist does. He, too, makes many measurements and averages them for more accurate results. He keeps complete records of what his measurements mean. He uses formulas to help him determine beforehand what he can expect the results to be. He repeats experiments of other physicists to verify their results.

What should happen if you increase the weight of the jar? According to the formula, it should roll farther. E = HW. If you increase the weight, you will increase the energy, and that will increase the length of the roll. Try it, using a heavier jar. If you haven't got a heavier one, try a lighter one. According to the formula, the roll should be shorter.

Now try this: fill the first jar you used with water. This adds considerably to its weight. According to the formula, this should make it go farther. Now try rolling it down the incline just as you did when it was empty. To your surprise, it will stop short of your previous marks! Is the formula incorrect? What has happened?

Remember, at the beginning of the experiment you assumed the friction of the air and of the rug to be constant for each roll. Now, however, you have added water, and the water can move around inside the jar as it rolls down the incline. The water can also move against the inside of the jar as it rolls over and over. This produces a tumbling action and the water "rubs" against the inside of the jar. At the back, for example, the side of the jar is going upward while the water is being pulled downward by gravity:

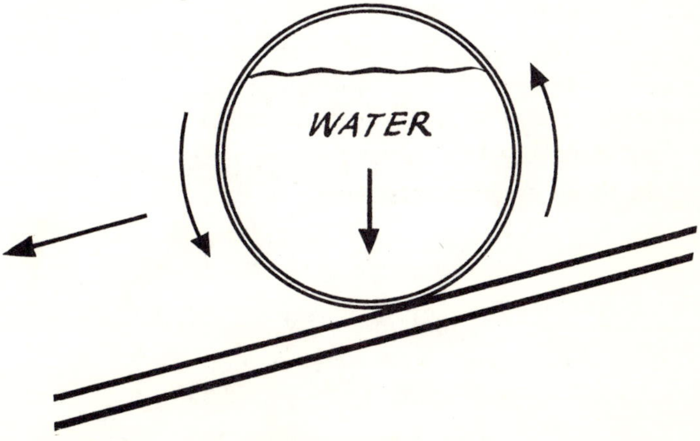

This friction continues as long as the jar rolls. It takes energy to overcome this added friction, and this causes the jar to stop short of its earlier marks. This is a difficult thing to believe unless you try it. By all means do.

As soon as you change the conditions of the experiment, you must be prepared for changes in the results. This is one of the reasons why an experimenter tries to keep every aspect of the conditions constant except the one he has under observation.

As you experimented, you were verifying the formula which had been worked out by other physicists. When your results were recorded, the formula helped explain why they came out the way they did. This is an important part of the work physicists do. They explain the basic how and why of such things as pressure cookers, television sets, cameras, nuclear energy, jet planes, echoes, microscopes, and electricity. Physicists try to understand these things as completely as they can. This means they continue to ask themselves "why?" until they finally come to the very nature of matter and energy.

Albert Einstein (1879–1955) was a physicist who tried to explain how new ideas about time, space, mass, motion, and gravity were related. His theory of relativity is one of the greatest intellectual achievements in the history of mankind. His famous equation: $E = MC^2$ was used to work out some of the problems relating to atomic energy. In his formula, E, the energy in ergs (a unit of energy used by scientists), is equal to the mass (in grams) times the speed of light (in centimeters per second) squared. While his formula looks simple, the brilliant thinking that was necessary to discover it is of the highest order.

Besides explaining the how and why of everyday things, a physicist primarily investigates phenomena not so closely related to our immediate lives. He studies cosmic rays, subatomic particles, ultra high-frequency sound, the structure of molecules and atoms, the behavior of light, why a magnet attracts only certain materials, the relationship of matter to energy, and many other things. The physicist in these in-

Here physicists and their assistants are preparing special apparatus for an experiment involving highly radioactive material. Note the protective clothing worn by the scientist on the left. (Photograph courtesy of Brookhaven National Laboratory.)

stances is more interested in finding out why something happens rather than to what use it can be put. This "pure" research has been responsible for some of the most important developments of our time.

With the growth of atomic power, the exploration of space, the investigation of the structure within the atoms, the development of new materials with more useful physical properties, physicists are today more sought after than ever before in history.

10. Collector's Items

Geologist

The next time you go on a hike, why not become a geologist? You may find that before long you will be taking hikes *because* you are a geologist. A geologist is a scientist who studies the earth's crust. You, as a beginning geologist, can start by making a collection of rocks and minerals taken from the earth's crust.

Why should you make a collection of rocks? Well, if you collect rocks and minerals, you'll get to know something of the composition of the earth's crust and learn about the resources that are found in the earth and how man uses them. You may even find some fossil material and go on to learn something about the plant and animal life of long ago.

The tools that you will need to collect rock and mineral specimens are relatively simple.

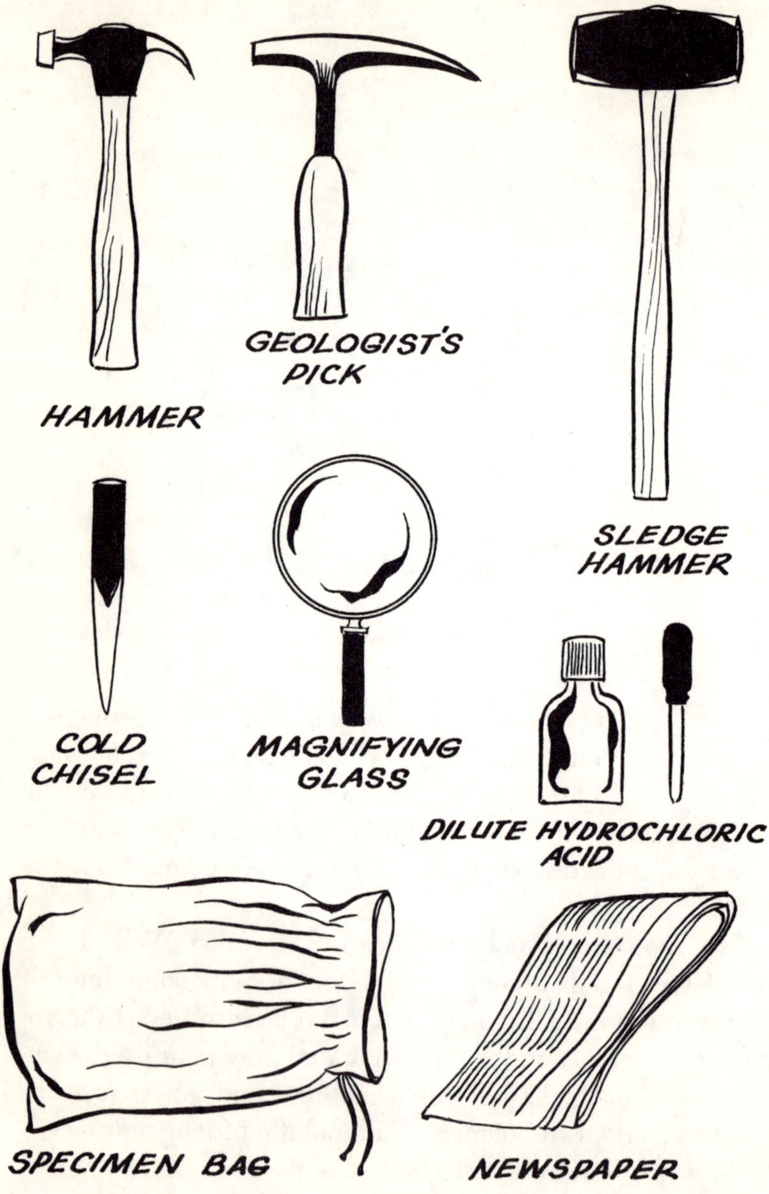
HAMMER
GEOLOGIST'S PICK
SLEDGE HAMMER
COLD CHISEL
MAGNIFYING GLASS
DILUTE HYDROCHLORIC ACID
SPECIMEN BAG
NEWSPAPER

An ordinary hammer can be used, but a geologist's hammer, or pick, is more useful. It looks like a regular hammer on one end but has a pointed end on the other. The hammer is used to remove the outer layer of rock which may have been changed by exposure to rain and air. It's also used to break off small pieces as you examine the rock and to get the piece in the proper shape for your collection. A small cold chisel, made of strong steel, is also useful in shaping the rock to the proper size. A magnifying glass is useful for examining the structure of the rock or mineral closely. A bag, box, or knapsack will be needed to hold the specimens, as well as a newspaper to wrap them in.

You can find specimens anywhere. In your hiking, picnicking or traveling across the country, keep looking for specimens. What should you look for? Well, in the beginning you will probably collect only those specimens which are attractive in shape, especially lustrous or, perhaps, transparent. Later on, as you get to know more about what to collect, you will choose specimens that show special veins of material and specimens which are valuable because they are rare. As you progress with your hobby, you will perhaps want to go to places where the chances of getting specimens are better, rather than just casually picking them up. The best place to go is where someone else has already begun to dig, or where nature has helped you in the digging.

Look for specimens where someone is excavating a house, or building, or where a new road is being made. Road builders often have to cut through hills and will then lay bare all sorts of rock specimens. A stone quarry is another place to look. Even though the men working there are looking for stone, they will probably be looking for stone of a different kind than you are. Another place to look is a stream bed. The younger, swifter rivers often uncover material for you. It's easy to see how a river can do this if you do a simple experiment with the following materials:

Stones
Pebbles
Sand
Soil
Water

On a relatively large, sloping surface such as a driveway, pavement, or a wide piece of wood, put a mixture of large and small stones, sand, and soil. Mix the ingredients thoroughly and keep them in a pile on top of the incline.

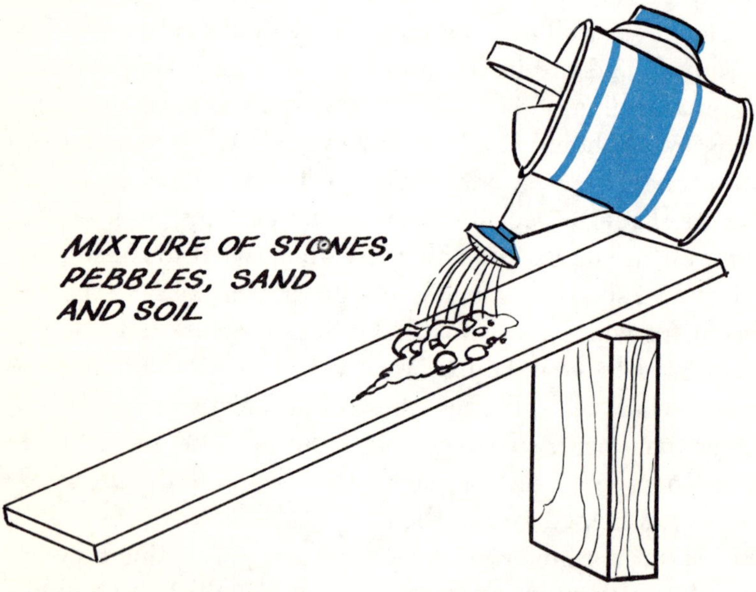

The pavement, driveway, or piece of wood should not be so steep that the mixture you place at the top begins to slide down. Now, slowly pour water onto the pile of stones, pebbles, sand, and soil. Continue to pour, and you'll see how a stream will help you in your hunt for geological specimens.

The water, as it runs down the incline, carries some of the material with it. Notice that the running water carries the soil away with it quite easily. Pebbles and sand wash down part of the way, but the larger stones probably remain behind.

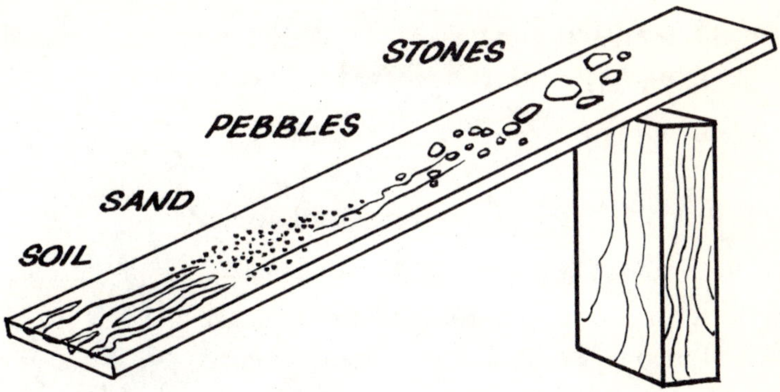

From this experiment, you can see how running water helps break up the materials of the soil and sorts them according to size. Water can carry solid material along with it until it begins to slow down. It deposits the heavy material, like rocks and stones, first; then the lighter material, such as pebbles, then sand, and, last of all, the soil. Therefore, when you look for streams that have done your digging for you, look for young streams with swiftly flowing water. There you'll find the most rocks. As the stream gets bigger and wider, the water slows down and begins to deposit more sand and silt which covers the rocks.

A geologist classifies rocks into three main types. *Igneous* rocks are rocks that were formed by fire, probably volcanic. Basalt is an igneous rock that's very common. It is a dark brown, green, gray, or black rock.

Sedimentary rocks have been formed out of material which has been deposited by water. Examples of sedimentary rock are: sandstone, which is merely grains of sand "glued" together; shale, which is a form of clay that has been pressed into rocks; and limestone, which is made up of small shells, or the mineral calcium carbonate, often mixed with sand or clay.

To see how sedimentary rocks are formed, you can do another experiment. Here's what you'll need:

- Quart jar with cover
- Sand
- Soil
- Pebbles
- Water
- Tablespoon

Put four tablespoons of sand, four tablespoons of fine soil, four tablespoons of fine pebbles into a quart jar. Fill the jar with water, put the top on, and shake vigorously. When you put down the jar on the table, watch closely what happens. The pebbles will go to the bottom immediately, then the sand will fall rapidly to form a layer, then a combination of sand and larger pieces from the soil, and finally a fine layer of material will be deposited on the top. If you do not move the jar for several days, the solid material will settle to the bottom in a fine layer and the water may be almost clear. See illustration on facing page.

This is the same thing that happens to material that is relatively light in weight and is carried by water to lakes or the oceans. As the water slows down, this material begins to settle in the same way as the material in the jar settled. The heavier pieces will be deposited first, then the lighter ones on top, so that layers are formed. If tremendous pressure or heat, or both, is applied to these layers, they can become hardened to rock.

Sometimes the water that helped form the layers has long

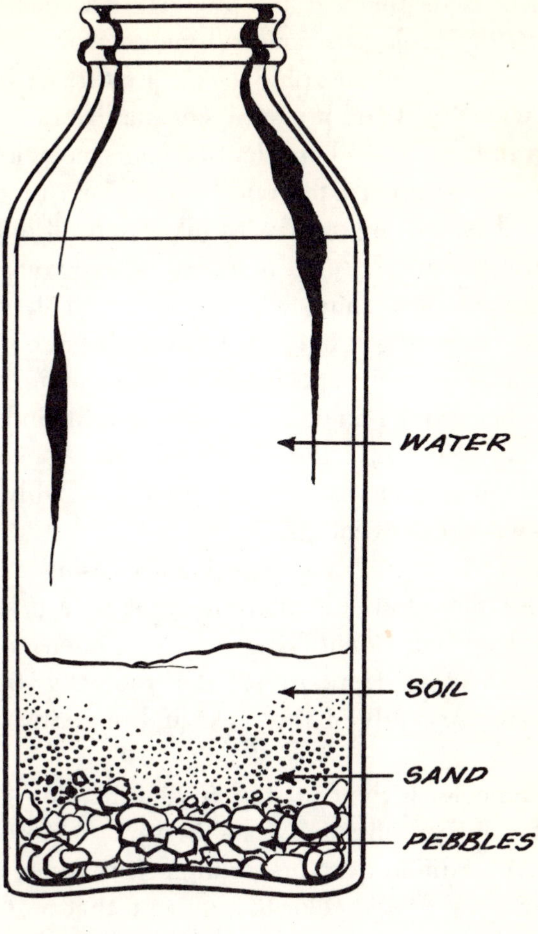

since disappeared because the land was raised. Today the layers that formerly were under a lake or ocean are now part of a hill. You often see these layers when a road crew cuts through rolling country.

The third kind of rock is *metamorphic*. These are rocks that have changed form or shape because of great heat and pressure inside the earth. Often they began as sedimentary rock but because of the heat and pressure were changed into rocks which the geologists now call metamorphic. Marble is a good example of metamorphic rock in that it was originally limestone which heat and pressure recrystallized into marble.

You certainly will want to identify your specimens when you get home. Number each one by either pasting a small number on it or writing on it with ink. Then fill out a small card or enter the number in a notebook where you will keep the information concerning where you found it, what the surrounding material was like, and so forth. Be sure to include the date. Look up your specimens in one of the reference books for rocks and minerals which you will find in your public library. No doubt the library will have several such volumes which you can examine. You will probably want to buy your own books eventually.

Sometimes identifying your specimen is going to be difficult because thousands of different minerals are known to geologists. However, identifying your specimens is part of the fun of collecting. Here are some of the tests you can do to help with your study of the rocks and minerals you have collected.

Noting the color is the first step in identification, but color alone is not always reliable. Often the same mineral can be found in several different colors, or impurities can change the color of a mineral or add color to a mineral that might otherwise be colorless.

The shine or luster of a mineral is also useful in identifying it. The way a mineral reflects the light is known as luster and can be described as earthy, metallic, waxy, glassy, greasy, dull, and so forth.

Often you will want to break the specimen apart to see what it is like inside. The break could shatter the piece like glass. It could be a smooth break or ragged break. These characteristics will help you identify the mineral. The sides of the break will also give you some indication of the crystal structure of the specimen. Here is where the magnifying glass will be useful.

Strangely enough, the mineral itself may be one color, but in powdered form can be quite different. Once powdered, however, the color is almost always the same. That's why geologists use the powder to help determine what the specimen is. You can try powdering bits of your specimens by rubbing them on a rough, hard, white surface like a piece of rough tile. If your specimen happens to be iron pyrite, for example, the crystals inside might be yellow, but when you rub them on the rough tile the powder will be greenish black.

Another characteristic of minerals and rocks is their hardness. A geologist rubs one mineral with another to find out which one becomes scratched. A scale has been developed so that the hardness can be referred to by numbers. See illustration on page 140.

After consulting your book of rocks and minerals and trying the tests, you still may not know for sure what the specimen is. You will have to try some chemical tests. If you are going to study your specimens at all seriously, you will need a small bottle of hydrochloric acid. You can probably get this from your teacher at school or, perhaps, at the drugstore. Explain what you need it for and the teacher or druggist will make you up a small bottle of the dilute acid. To make the test, simply put one drop of the acid on a corner of the specimen. If it bubbles, you know that the specimen has calcium carbonate in it. This means that it is probably marble, limestone, or chalk. The reason the bubbling takes place is because the hydrochloric acid reacts with the calcium carbonate, and carbon dioxide gas is given off.

From your investigations of books about geology and

HARDNESS	TEST	EXAMPLE
1	SCRATCHES EASILY WITH FINGERNAIL	TALC
2	SCRATCHES WITH FINGERNAIL	GYPSUM
3	SCRATCHES WITH PIN	CALCITE
4	SCRATCHES EASILY WITH KNIFE	FLUORITE
5	SCRATCHES WITH KNIFE	APATITE
6	KNIFE WILL NOT SCRATCH WILL SCRATCH GLASS	FELDSPAR
7	SCRATCHES GLASS EASILY	QUARTZ
8	SCRATCHES QUARTZ EASILY	TOPAZ
9	SCRATCHES TOPAZ EASILY	CORUNDUM
10	SCRATCHES ALL OTHERS SCRATCHES OTHER DIAMONDS	DIAMOND

mineral collecting you will discover that the geologist has still other tests. In some he uses a special blowpipe to melt down specimens to see if they contain minerals, and at the same time he can check to see what kind of gases are given off.

He can also dissolve some of them in various liquids, or at least try to, and thus get information about them.

It's possible that as you are looking for specimens you may find some that are relatively valuable as gems. Gems are simply minerals that people consider valuable because of their beauty. Agate is found in a great variety of colors and frequently has a translucent quality that is very attractive. Azurite is deep blue, opaque and waxy. Dinosaur bones will often be black, brown, or red and are usually opaque and waxy. If you have a chance to collect specimens in Oregon or California you may discover some American jade. This has various colors, ranging from white to green, is glassy and translucent. However, you don't have much chance of discovering more precious stones like zircons or diamonds, emeralds or rubies.

You may want to investigate the possibility of forming some of your specimens into jewelry by cutting or polishing them. A person who does this is known as a lapidary. It is not easy, but it's a good companion hobby to collecting geological specimens.

The geologist, aside from collecting geological specimens, uses the specimens to find out information about the earth's crust. With his experience and knowledge he can often discover what the conditions of the earth in a certain vicinity were like billions of years ago. He can also do this by studying the layers of material in the ground. Fossil material that you may discover will probably be shell outlines or formations embedded in rock. Treat these carefully, because they make attractive additions to your collection.

After a while you will want to do more than just collect specimens. Try concentrating on the geology of your local area. Make a special collection of specimens that come from nearby, and try to work out what they mean in terms of the

A geologist and his assistant making measurements near Mackay, Idaho, as part of a survey of an area that holds promise for development of oil. (Photograph courtesy of Standard Oil Co. [N. J.])

geology of the area. You can even develop the history of the land as told by the specimens you collect.

As you studied your collection and got to know more and more about rocks and minerals, you were gradually becoming expert in one branch of geology. Other geologists specialize in volcanoes, glaciers, ocean bottoms, fossils, the physics of the earth, the chemistry of the earth.

Trained geologists are employed by companies that use raw materials that come from the ground. Oil, coal, metals, gems, sulphur, asbestos, borax, salt, water all come from the ground and are carefully investigated by geologists. Individual states and the federal government need geologists to make geological surveys and investigate specific areas.

Colleges and universities have geologists on their staffs to teach geology. They often act as consultants because of their specialized knowledge.

Perhaps your interest in collecting rocks will lead you to find out more about one of the branches of geology. You may even discover that you would like to become a specialist in one phase of geology. In any case, being an amateur geologist gives you a chance to collect, perhaps, something valuable.

It helps you exercise out in the open. It can lead you on to some real adventures, and, most of all, it can be a lot of fun.

11. Blowing Up Balloons with Yeast

Biochemist

In the following experiments you're going to be investigating the chemistry of life. In doing so you're going to be a biochemist, short for "biological chemist," a scientist who searches after the chemical secrets of living things. The living things you'll be working with are yeast plants, and they're going to blow up balloons!

A cake of yeast, or powdered yeast, doesn't look very much alive, but it is, just the same. A cake of yeast is really made up of thousands of tiny yeast plants that need only the right conditions to begin to grow and multiply. You're going to grow several colonies of yeast plants by supplying them with identical conditions except for one. You're going to give each colony a different kind of food. Which food will be the best for the yeast? That's one of the things you're going to discover for yourself.

You'll also discover, as you work in the field of biochemistry, that there are many similarities between yeast plants and you!

Collect the necessary equipment and "chemicals" from the kitchen and perhaps the grocery store. If you can't find them all, use what you can find and even make substitutions. Just keep track of what you use. Here are the suggested materials:

Yeast	4 large balloons
Syrup	2 glasses
Flour	4 quart bottles
Gelatin	Funnel
Grape juice	String
Tablespoon	Notebook and pencil

First, number the bottles with a grease pencil or with numbers penciled on pieces of paper which you can glue or paste to the bottles. As you prepare the contents of each bottle, make a note of what and how much went into it.

In one of the glasses, make up the yeast solution. You can get yeast either in the dry or cake form. Follow the directions on the package, but put three packages or cakes into one cup of warm water.

Try to use all the same kind of quart bottles if you can. The kind that soda water comes in works fine. In the second glass put a half cup of corn syrup and four tablespoons of the yeast solution. Mix thoroughly and, through the funnel, pour into the first bottle. Label the bottle "Corn Syrup" and note the time and date.

Wash out the glass and funnel. Next, mix a half cup of flour and a half cup of water in the glass and add four tablespoons of the yeast solution. Mix well and put into the second bottle. Label.

Wash out the glass and funnel again and make up the gelatin solution according to the directions on the package, which will probably be one tablespoon of powder to a half cup of water. Add four tablespoons of the yeast solution and pour into the third bottle. Label.

Wash the glass and funnel and pour out a half cup of grape juice, add four tablespoons of yeast solution, and stir. Pour this into bottle 4 and label.

Slip a balloon over the mouth of each of the bottles and tie a string around the neck of the bottle firmly so nothing can get in or out of the bottle. Lay all the bottles on their sides in as warm a place as you can find (about 85° F. is ideal). They will now look like this:

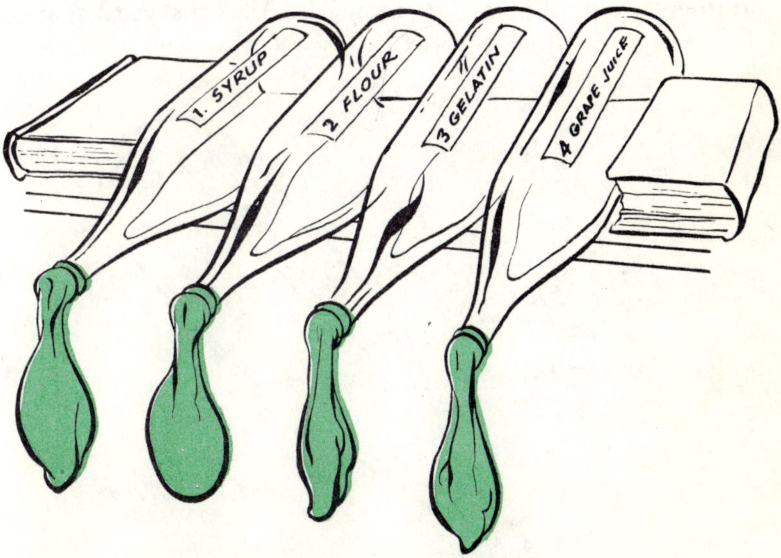

The foods that you've supplied to the yeast plants represent different types of foods. The syrup is mostly sugar that has been partially processed. The flour is high in starch. The gelatin is a good example of a high protein food. The grape juice has a high percentage of natural sugar in it. All of them provide food that's necessary for your growth. Which one will be the best food for the yeast?

It's fun to guess what's going to happen. You might even make a list of the various foods, starting with the food you think will be the best and ending with what you believe to be the worst. Your family and friends can join you in trying to predict the results. Write down all the guesses and compare them with the actual results as the experiment progresses.

In a half an hour you will see changes taking place in some of the bottles. Be sure to record the time, temperature, and condition of each of the bottles and their balloons whenever you make your observations. In some, small bubbles will be forming. The bubbles are an indication that the yeast plants have started to become active:

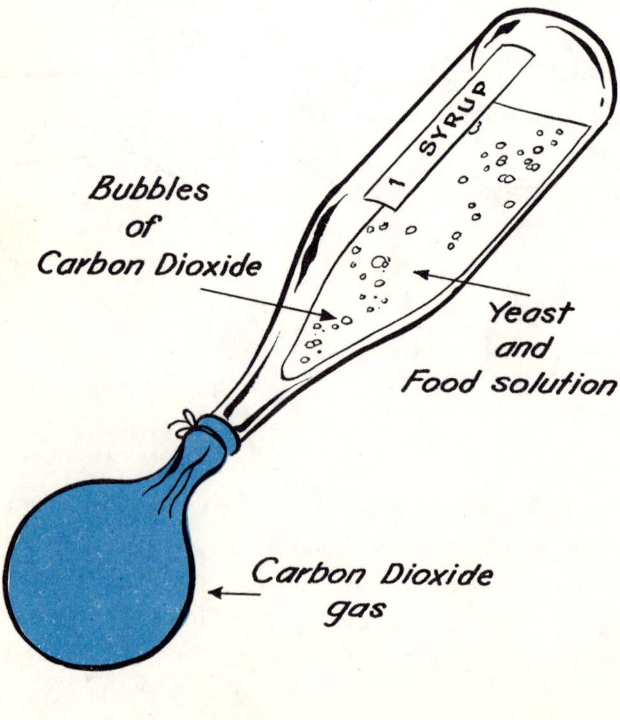

About every half hour or so, turn each bottle upright and swirl the liquid inside to mix it as thoroughly as you can. This brings new food to the yeast plants and will help them grow.

The tests you have begun are the sort of experiments early scientists conducted with yeast. A French scientist, Charles Cagniard de la Tour, first saw yeast cells reproducing or "budding" under his microscope in 1857. He also noted that when the yeast plants acted on sugar in his solutions, they formed alcohol. Later, Louis Pasteur proved that the living yeast cells produced the alcohol by a process known as fermentation. Pasteur performed experiments like these you are doing and from them he showed that every kind of fermentation is caused by a living thing. Most of the time the living things are too small to be seen except with a microscope.

If you have made the microscope described in one of the other chapters, use it to see the yeast cells that are growing in your test bottles. Simply make a very watery mixture of the yeast solution and place it under the microscope. The yeast plants will appear as rounded bodies. You might even see one with a bump on one side. If you do, watch it for a while. It may be budding, and you may actually see the plant separate into two plants. This is the budding process Cagniard de la Tour first observed.

The yeast plants in your bottles need food, water, air, and the proper temperature to grow. Check your notes and you'll see that you put about the same amount of water into each bottle. All the bottles have been exposed to the same amount of air in the bottles and balloons. They've also been at the same temperature. The only important variable has been the kind of food you supplied to the yeast.

As each yeast plant grows, it forms chemicals called enzymes. These enzymes break down the sugar and starch in the various solutions if the temperature is 80–85° F. and the proper amount of moisture is present.

When yeast is added to bread dough, the yeast cells produce the enzymes that change the starch in the flour to sugar. The sugar is acted upon further to change it to alcohol. In the process, carbon dioxide gas is given off. This carbon dioxide gas is inside the bubbles in your various bottles.

With this knowledge, you can now determine which bottle contains the best food for the growth of yeast. The bubbles will expand the balloons, so the bottle with the largest balloon will be the bottle which contains the best food for yeast.

You might be wondering what becomes of the alcohol. It is an important part of the final product when various yeasts are used to ferment prepared grape juices in the making of wines. The alcohol content of beer and its characteristic flavor also are results of the special yeast used in the fermentation of the brew. You should be able to detect the presence of alcohol in any of your bottles by the smell. One may even smell like wine.

The alcohol formed when yeast makes bread dough rise is driven off by the heat when the bread is baked. There's no alcohol in the bread when you eat it. The yeast is used to make bread lighter, easier to digest, and better tasting.

The alcohol made with the help of yeast is extremely important in the chemical industry. Alcohol is used to help make shellac, varnish, soap, celluloid, gunpowder, drugs, perfumes, and hundreds of other materials. Almost all of the alcohol used in industry is made by the same method that you made alcohol: the "fermentation" by yeast.

Not only is alcohol valuable as an industrial product, but it is also important because it is a poison. Alcohol is often used to kill the lower forms of life that we want to control. At the same time it is a poison to humans as well. When drunk, it slowly acts to poison the body.

After an hour or two have gone by, you should easily be able to tell, by the size of the balloons, which of the bottles contains the best food for the yeast. Biochemists who are in charge of growing yeast commercially judge the various strains by measuring the amount of carbon dioxide gas they release, just as you are doing. As a good biochemist, however, you would probably conduct many more experiments, not only to verify your results, but also to try many more kinds of food before you would decide which is best.

After this test is finished you might like to try other materials to see how the yeast plants thrive on them. You can try any of the following, or think up some of your own: milk, tomato juice, soup, cornstarch, molasses, sugar solution, and vegetable oil.

When you make up new solutions, try to get the same consistency as in the first test mixtures. Add the yeast and tie on the balloons. Remember, yeast plants need more than just energy-producing foods to grow. In this way they are much like you are. You wouldn't stay healthy very long on nothing but sweets because you need a well-balanced supply of food. So do the yeast plants. The food you supplied may not have had the proper minerals which yeast plants need to grow. A source of nitrogen is important to them also.

Now that you understand how the experiment is set up, you can prepare as many bottles as you want. Just make sure they are all the quart size, and keep accurate records for each bottle.

What chemicals will be harmful to yeast plants? This information would also be important for the biochemist to know. To see how he would go about finding out, you can do another series of experiments.

Get together the following "chemicals":

Yeast
The best food from your previous tests
Vinegar
Ammonia
Laundry bleach
2 glasses
Teaspoon
4 bottles
Tablespoon
4 balloons
Funnel
Notebook and pencil

Make up a cup and a quarter of the mixture that you found to be the best food for the yeast plants in your previous experiment. Put a quarter of a cup of this food into each of the bottles. Next, make up the yeast solution as you did before. Add two tablespoons of the yeast solution to each of the bottles as before and stir. Label the bottles as before and keep a record of what you put into each bottle.

Prepare the four bottles like this:

Bottle 1. Add nothing to the food and yeast that's already in the bottle. This bottle is your "control" and will be used later to compare to the other bottles.

Bottle 2. To the food and yeast, add one tablespoon of vinegar, and stir.

Bottle 3. To the food and yeast, add one tablespoon of ammonia, and stir.

Bottle 4. To the food and yeast, add one tablespoon of bleach, and stir.

Tie the balloons in place and put the bottles in the same warm place you did before.

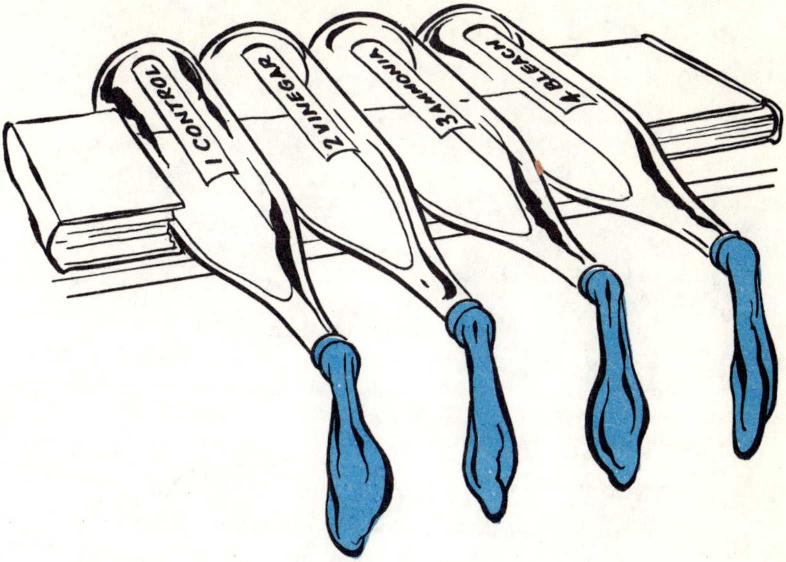

Check the progress of the yeast growth every half hour or so. Be sure to record the time, temperature, and condition of the solutions and the balloons every time you make an observation, then twirl the bottles as before to mix the contents.

Some of the chemicals you added to the food and yeast are poisons to the yeast plants. You will be able to tell which bottles contain the poisons, because very few or perhaps no bubbles at all will form and the balloons will not expand. Not all of the materials are poisons however. You may be surprised at some of your results.

If you'd like to try other household materials to see how yeast plants react to them, here are some suggestions: rubbing alcohol, turpentine, milk, wine, tea, coffee.

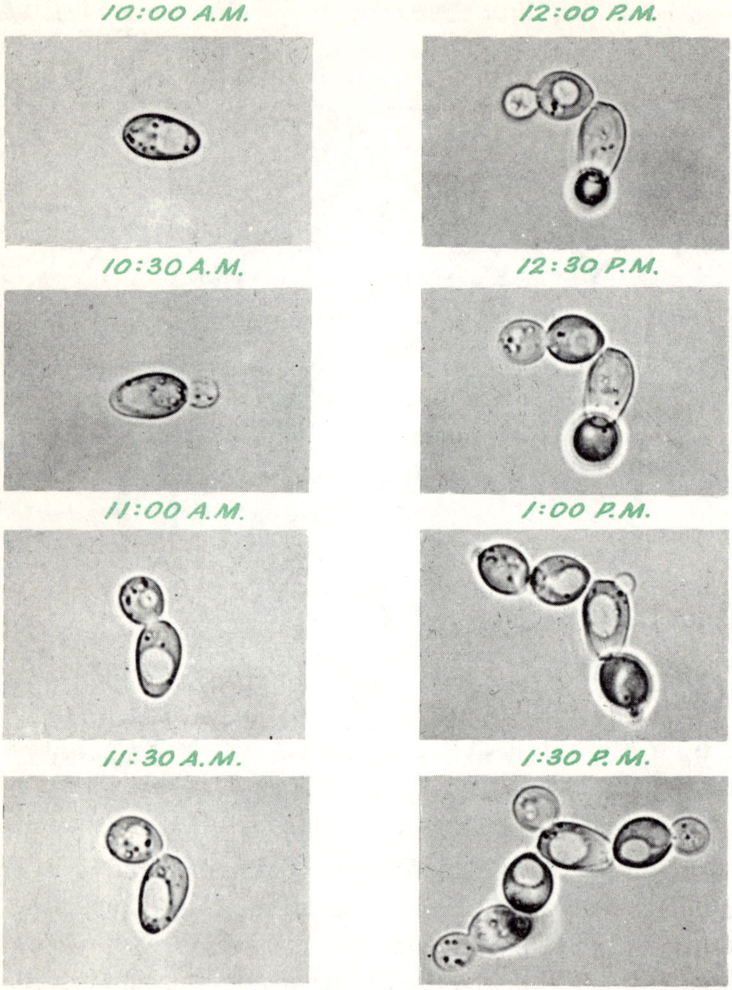

Yeast cells magnified 1000 times are shown at half-hour intervals. Note that because the biochemist has supplied the yeast with the proper conditions, one yeast plant has become two in about an hour or so. (Photograph courtesy of Fleischmann Yeast Co.)

Make up bottles of your best food as you did the first time, even though you may have since found a better food. If you use the new food, you won't be able to compare the results you get this time with what you got last time. Scientists try to limit the things that can be varied from one experiment to the next. In all your experiments, wherever possible, change only one thing at a time.

To your "standard stock" solution, add the yeast and then the substance you want to test. Record the diameter of the balloon if and when the yeast forms carbon dioxide.

You can also try growing yeast plants in your best "medium" and varying the amount of test material. For example, you could make up four bottles (or more if you have them) each with syrup, water and yeast solution. Then into the first, put a quarter of a teaspoon of rubbing alcohol, in the next, a half teaspoon, then three quarters of a teaspoon and then one teaspoon . . . and so on. This way you'll be finding out what quantity of the materials makes a difference to the growth of the yeast.

Louis Pasteur found out that lactic and vinegar bacilli, or bacteria, produce acids that keep yeast from growing and sour the fermenting liquid. Lactic acid is what sours milk. When you added vinegar (acetic acid) to your best mixture for growing yeast, what happened? What do you think would happen if you tried sour milk?

Biochemists work to improve our food supply by improving materials needed to grow food and developing others to prevent decay and spoilage. Biochemists discovered vitamins, the chemicals needed by our bodies to stay healthy. Other biochemists discovered ways of manufacturing them in quantity. Still other biochemists investigate the chemistry of life just to find out more about it. They often teach in universities but spend part of each week trying to solve some of the mysteries of life.

These research scientists have discovered that you have things in common with a cake of yeast. Your muscles produce enzymes just as yeast does. When you use your muscles, the enzymes help a process of fermentation take place. Muscles which lack the enzymes, because of injury, can be restored to health by supplying enzymes from yeast!

12. A Detective and His Solutions

Analytical Chemist

Here's your chance to be a chemical detective. Show a friend of yours four quart bottles containing clear liquids. Explain that you're going to leave the room and he is to put any combination of the liquids from the four bottles into five glasses. When you return, you're going to tell him exactly which liquids he put into each glass.

Your friend may think you are a magician. You can tell him, however, that you're an analytical chemist, because you'll be doing the same kind of chemical detective work that such a scientist does.

The necessary materials are all quite ordinary for results that are so startling.

Lemon juice	Bottles
Tincture of iodine	Teaspoons
Cold-water starch	Measuring cup
Purple cabbage	Note book
Glasses	Pencil

First, make up what will be known as the stock solutions.

Stock Water Solution: One quart of water.

Stock Starch Solution: Into a quart of warm water dissolve a quarter teaspoon instant laundry starch.

Stock Lemon Juice Solution: Into a quart of water stir the juice of one lemon or four teaspoons of bottled lemon juice.

Stock Sodium Bicarbonate Solution: Into a quart of water stir four level teaspoons of sodium bicarbonate.

When you have finished, all four quarts of liquid will be clear. Label them in order to tell which is which.

Next, make up the indicator solutions.

Base Indicator: Simply pour off the juice from a can of cooked red cabbage. Fresh red cabbage can be used by chopping some of the leaves into very small pieces. Put them into a saucepan, cover them with water, bring them to a boil, and let them simmer for about twenty minutes or until the liquid is dark red. While it's cooking, make out a label that says: "Base Indicator–Base: blue." When the liquid is cool, pour it into a bottle, cap it, and paste on the label.

Starch Indicator: Go to the medicine cabinet and get the tincture of iodine. This will very likely be a 2 per cent solution of iodine dissolved in alcohol. Use it just as it is.

Vitamin C Indicator: Pour one cup of the stock starch solution you made into a smaller bottle. To it, add about ten drops of tincture of iodine, or add until it is a deep blue color. Label it "Vitamin C Indicator."

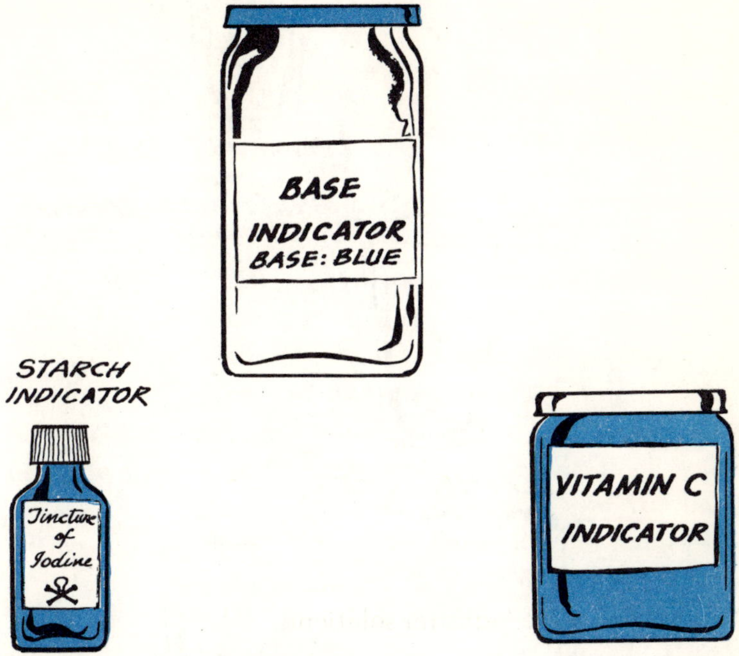

You are now ready to explain to your friend what he is supposed to do. Present him with the stock solutions and tell him he can put any combination of these solutions together in five glasses. He must use a quarter of a cup of the solution or solutions he chooses. He can add water to any glass to bring its level to match those in the other glasses if he likes. You can turn your back as he prepares the glasses, or, better yet, leave the room.

Let's say he mixes the solutions in the five glasses like this:

It's a good idea to supply your friend with a chart for keeping track of what he's put into each of the glasses. The chart could look something like the one that follows, and on it your friend might have made checks like those shown on page 162.

	STARCH blue	SODIUM BICARB. blue	VIT. C clear	WATER clear
1				✓
2	✓	✓	✓	
3			✓	✓
4	✓			✓
5		✓		✓

When you return to the room, the glasses of liquid will look exactly alike, and trying to find out what went into each one looks like a real problem. And it is—to anyone who is not aware of the methods of the analytical chemist. But you have prepared indicators and can now put them to good use. They will "indicate" to you what went into each of the glasses by a color change.

Line the three little glasses in front of the first mystery glass. Put two teaspoons of the liquid from glass 1 into each of the smaller glasses. These are your test samples. Into the first test glass, put a few drops of tincture of iodine. If the liquid in the glass turns dark blue, it has starch in it. If it remains clear or perhaps slightly yellowed from the iodine, there is no starch in it. The sample you're testing shows no starch present, because there is no change to a blue color.

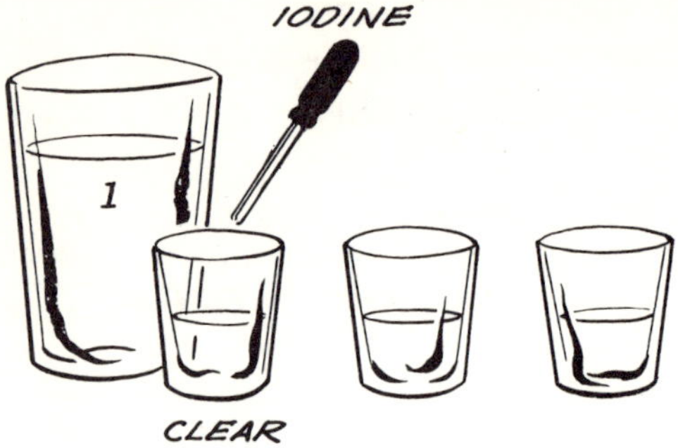

You know by now that a good scientist always records his results. As an analytical chemist, it is especially important that you record your results; otherwise you will forget what the tests show and will become very confused by the time you have tested all five glasses. Your notes could take the form of the chart you prepared for your friend. The next illustration will help you make up yours. Note the colors indicated at the top of the columns right under the name of the indicator. These will be explained as you make each test.

Right now you've finished the first test and can write "clear" in the first square in the starch column. The word "blue" at the top of that column indicates that if a blue color results, starch is present.

	STARCH blue	SODIUM BICARB. blue	VIT. C clear	WATER clear
1	clear			
2				

Into the next small glass put one teaspoon of the base indicator. If it turns blue, your friend put sodium bicarbonate into the glass. If it stays red, there is no sodium bicarbonate in the glass. In the example, the indicator stays red, showing no sodium bicarbonate in the glass.

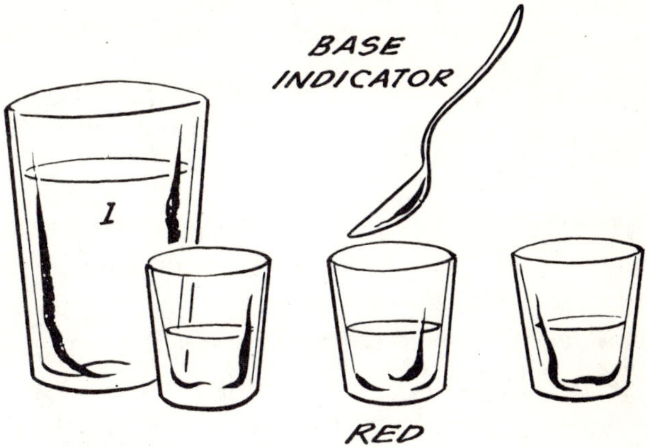

On the chart you can write "red" in the second column under "base." Note the word "blue" at the top of the column. This tells you the color that shows sodium bicarbonate is present.

	STARCH blue	SODIUM BICARB. blue	VIT. C clear	WATER clear
1	clear	red		
2				

Lemon juice contains vitamin C, and this is what will indicate the presence of lemon juice. Into the third sample glass put one teaspoon of the "Vitamin C Indicator." If it remains blue, there is no lemon in the glass. If it clears with no blue color remaining, there is lemon juice in the liquid. The example shows a blue color, indicating no lemon juice.

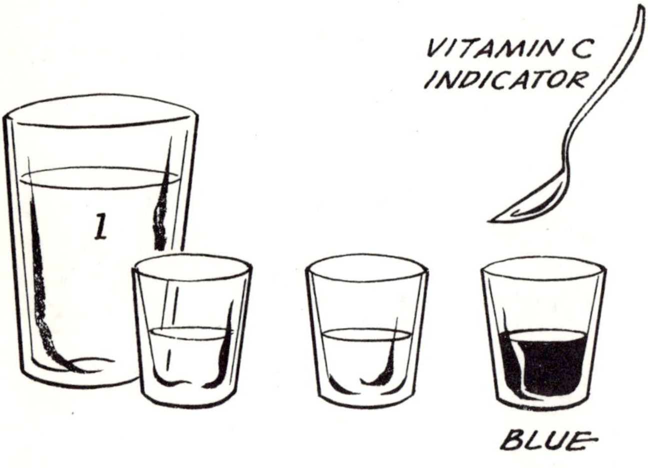

Write the word "blue" in the third column on your chart, and note the word "clear" at the top of the column.

	STARCH blue	SODIUM BICARB blue	VIT C clear	WATER clear
1	clear	red	blue	✓
2				

When you check your results you can see that none of the words you have written opposite glass 1 matches the words at the tops of the columns. This means the tests showed no starch, no sodium bicarbonate, and no lemon juice present. Your friend must have put nothing but water into the first glass. You can announce this to him. He can check his list and find that you are absolutely right!

Wash out the small test glasses and the spoon and repeat the tests with the second glass. Put two teaspoons of the liquid from the second glass into each of the test glasses. Add the proper amounts of the indicators as you did before.

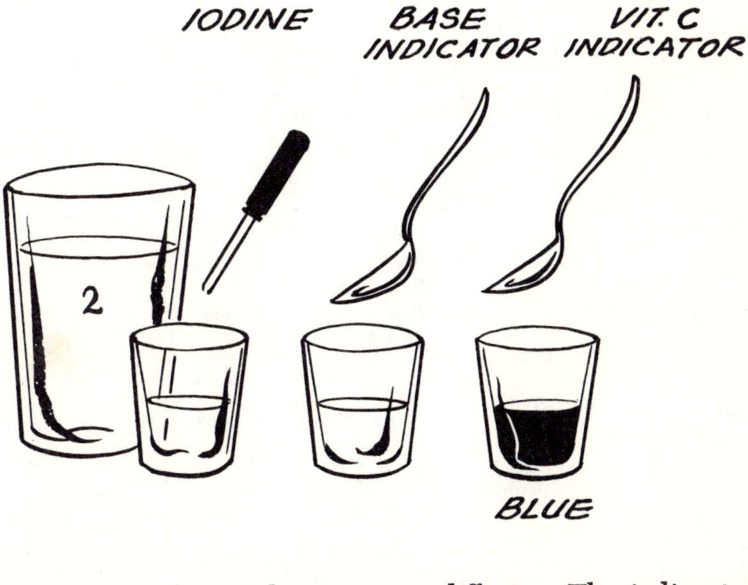

This time the results are quite different. The iodine test turns blue, showing starch is present. The base indicator turns blue, showing sodium bicarbonate has been added, and the vitamin C test turns clear, showing the presence of lemon juice. Your chart now looks like the one on page 166.

	STARCH blue	SODIUM BICARB. blue	VIT. C clear	WATER clear
1	clear	red	blue	✓
2	blue ✓	blue	clear ✓	
3				
4				
5				

Your friend had put all three of the stock solutions into glass 2. He will have to agree with you again.

You can continue to check the other glasses in the same way and record your results on the chart. When you finish, it will look like this:

	STARCH blue	SODIUM BICARB. blue	VIT. C clear	WATER clear
1	clear	red	blue	✓
2	blue ✓	blue ✓	clear ✓	
3	clear	red	clear ✓	✓
4	blue ✓	red	blue	✓
5	clear	blue ✓	blue	✓

If you look back to the list your friend made out, you will see that your chart matches it exactly.

You can see that, with the three indicators you've made, you can find out any combination of stock solutions your friend puts together. Just remind him that he must put a quarter of a cup of the stock solution he chooses into the glass. He can hide the stock solutions so that you can't see them while you're making the test. He might think that you can tell what he's used by the fact that the stock solution level has gone down.

If the mystery glasses have only a quarter of a cup of liquid in them, it means your friend put only one of the stock solutions into it. Of course half a glass indicated two solutions. Just remember the units can be any of the stock solutions including water. Thus, if you find only one of the chemicals present in a half a cup of liquid, your friend must have put in a quarter of a cup of water and a quarter of a cup of the indicated chemical.

Sometimes it's difficult to remember just exactly what indicator is supposed to indicate what, and how. Knowing more about what's going on should help you keep the reactions straight.

When you poured out the red cabbage juice, or boiled the fresh cabbage to get the juice, you were preparing a base indicator that works very much like the litmus paper that the analytical chemist uses. With it he can tell whether a solution is acid or base, because it, too, changes color. You can be sure sodium bicarbonate was present when your indicator turned blue, because sodium bicarbonate in water is a base. If the indicator remains red, this cannot indicate the presence of an acid, because the indicator is probably already acid. This is unfortunate, because lemon juice is an acid, and this test would help you uncover it.

A chemist wants to know if something is an acid, because

this gives him a lot of information about it. All acids are similar in that they have a sour taste, affect indicators in the same way, contain hydrogen, dissolve metals, and neutralize bases. Some acids are more active than others, of course.

Bases, on the other hand, are the chemical opposites of acids. They neutralize acids, combine with acids to form salts, have a bitter taste in solution, affect indicators all the same way (opposite of an acid), feel soapy, and contain metals combined with hydrogen and oxygen.

You might want to check other chemicals around the house with your base indicator. Try any of the following: ammonia, sugar, washing soda, borax. They should all turn the cabbage juice base indicator blue—except one. Which one?

When you have a good dark blue color by adding a base to the base indicator, try adding the stock lemon juice solution slowly and stirring. The acid in the lemon will neutralize the sodium bicarbonate and make the solution acid. The indicator will turn red! You can repeat with other liquids and thus test them to see if they are acids. Try vinegar, cream of tartar solution, and orange juice. Are all of them acids? Be sure to start with the blue solution by mixing the base indicator with sodium bicarbonate, then add the new liquid. If a red color results, it's an acid.

Other natural juices can also be used as indicators. You might try hollyhocks, red dahlia, red petunia, red iris, huckleberries, elderberries, and rhubarb juice. Boil and simmer in the same way you did the purple cabbage leaves. Then bottle, cool, and label. Test with sodium bicarbonate solution, and perhaps vinegar, to find out exactly how your new indicator will react. Perhaps you can discover some new indicators of your own.

Chemists have to deal with thousands of different chemicals. Classifying them into groups is often a help in finding out what they are and knowing what they will do when you mix them with other chemicals. In making up these groups, chemists look for similarities. Acids are one such group. Bases are another.

You used the iodine to test for starch. Long ago chemists discovered that when starch and iodine come together they form a deep blue color. Starch is the only material that will do this with iodine. So, if you know that one of the chemicals is iodine and it turns a second unknown chemical a dark blue, you can be sure it is starch. This starch-iodine test is one of the most reliable in the analytical chemist's list of tests.

When you added the dark "Vitamin C Indicator" to the various solutions, you knew, of course, that you weren't really testing for the lemon juice. The dark blue liquid you made by putting starch and iodine together becomes clear when mixed with vitamin C or "ascorbic acid" as the chemist calls it. Other chemicals will turn the iodine-starch solution clear also, but not baking soda. More starch will not change the color except perhaps to darken it. Therefore, of the three chemicals that can be in the glasses, only lemon juice will make the dark-blue indicator colorless.

You can see how the blue indicator will change by adding other liquids to it. Try dilute coffee, ammonia, dilute milk, grapefruit or orange juice, tea, etc. Remember, however, in these cases the color change will not necessarily indicate the presence of vitamin C.

An analytical chemist has many more tests than those you've been doing, of course. For example, when chemicals are combined with oxygen, the results, called "oxides," often do not dissolve in water. If these compounds are formed when two solutions are mixed together, the chemist can recognize them in the precipitate, or powder, that settles to the bottom. Color changes such as the starch-iodine test and the lemon juice test are very important to the analytical chemist. Other techniques of testing chemicals are: burning, applying various acids, analyzing the gases given off during a chemical reaction, applying electricity, and many others.

In these tests the chemist is often interested only in finding out *what* an unknown material is made up of without finding out *how much* of the material is present. This is known as "qualitative analysis." Other times the amounts of the chemicals are of vital importance. Very delicate procedures are used to determine the exact amount of each substance. This is known as "quantitative analysis." You did this in a very rough sort of way when you noticed how much water was in the glass you were analyzing.

The techniques that chemists have worked out to detect chemicals are amazing. For example, an ounce of the element lithium can be detected in more than nine tons of other material. Five parts of the element arsenic can be detected in 100, 000 parts of human hair.

The tools to do such delicate work are just as amazing. The sensitive analytical balance used by the chemist to weigh materials is a good example. Some balances are so precise that they can measure something that weighs only one one hundred fifty millionth of an ounce.

As you might imagine, analytical chemists have examined thousands and thousands of materials. There are specialists in analytical chemistry who will analyze anything you bring to them. Other chemists who investigate the make-up of materials teach chemistry and do research as well. One such chemist, and his co-workers at Princeton University, analyzed ancient bronze coins. They noted that the earlier the coin was made, the more tin and the less lead it contained. There was a more or less regular increase in the amount of lead and a decrease in the amount of tin as time passed. This gradual

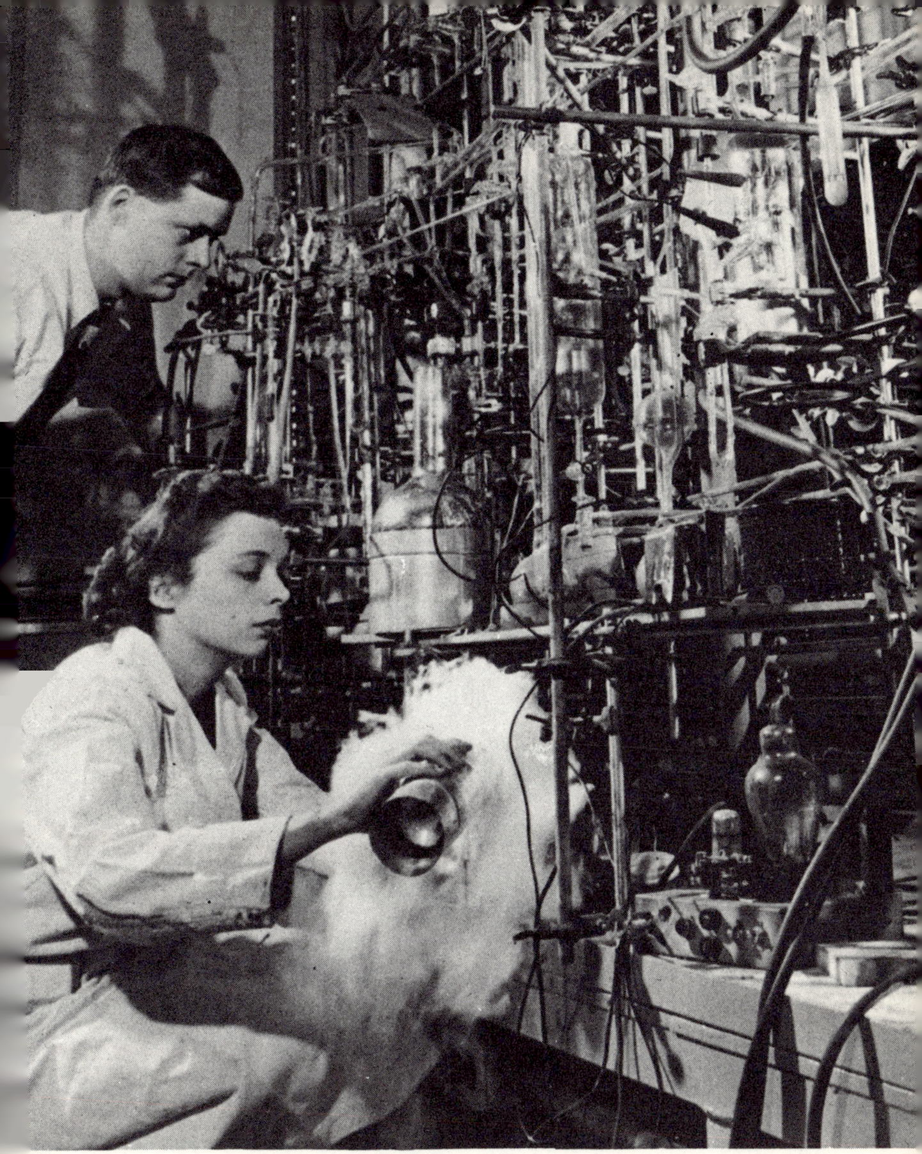

The analytical chemist sometimes uses elaborate and delicate equipment to find out what and how much of a chemical is present in a sample. (Photograph courtesy of Gulf Oil Co.)

change was probably due to the supply of tin running out while the supply of lead increased. With this information, the chemists helped the archaeologists establish a fairly accurate date for the making of undated coins.

The analytical chemist can be given the job of analyzing materials from anywhere on the earth to determine exactly what they contain. Sometimes the substance to be analyzed is used in connection with manufacturing processes. The analytical chemist can determine what per cent of certain materials is in the finished products and can help other chemists in their attempts to keep the materials pure.

You can see now why the analytical chemist is really a chemical detective.

13. Weighing the Air

Meteorologist

How would you like to become a meteorologist? If you've never heard the word before, you might guess that you'd be investigating meteors from outer space. But the Greek word "meteor" means "things in the air." As a meteorologist you'll be studying "things in the air." And when you know more about the "things in the air," you can try your hand at predicting what tomorrow's weather will be.

A meteorologist has many instruments to help him measure and record the "things in the air." One of the most important of his instruments is the barometer. It's a long tube closed at one end, filled with mercury, and turned upside down with its open end under more mercury in a dish. The height of the column of mercury varies with the pressure of the atmosphere. As you will see, the pressure of the atmosphere is important to know if you are going to predict the weather.

Unfortunately, the mercury is poisonous and difficult to get in sufficient amounts under ordinary circumstances. However, you can make a barometer that's perfectly safe out of a rubber sink plunger, some wood, and a piece of mirror. And, with it, you can try to forecast what the weather is going to be.

Here is a list of the materials:

Rubber sink plunger
Piece of glass
Piece of mirror
Dowel rod or construction set piece
Assorted wood
Pins
Paper
Pencil

The general appearance of the barometer can be seen here:

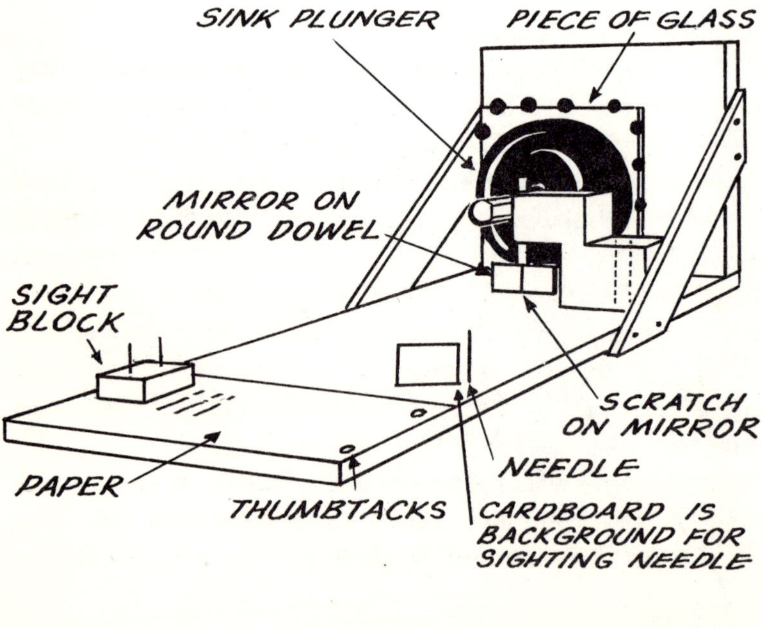

The barometer works because atmospheric pressure pushes against the rubber sink plunger. When you push the plunger into place, some of the air inside is forced out. The rubber is elastic and tries to spring back into its original shape. The air captured inside expands to fill up the increased volume of the plunger. The pressure of the air inside the plunger, plus the force due to the elasticity of the rubber, is equal to the outside air pressure. When the atmospheric pressure increases, the rubber is pushed in. When the air pressure gets less, the air inside pushes the plunger out. This movement back and forth is in direct proportion to the change in air pressure. If you can measure the amount of movement, you can relate it to the change in pressure.

Unfortunately, the amount of change is small. This means that some method of magnifying the movement is necessary before it can be measured. The magnification is accomplished with the help of a mirror. The mirror rotates as the plunger handle moves in and out with the changes in air pressure. See illustration on page 176.

The base and upright of the barometer can be made of any wood that's handy. Those shown in the drawings were made of half-inch plywood. The plunger is a four-inch diameter sink plunger carried at most hardware stores. A larger plunger can be used if the smaller version is not available. The height of the back is determined by the diameter of the plunger. The piece of glass is cut to fit the plunger and held in place with thumbtacks or nails with large heads. A support to hold the back is needed, because any movement of the back will be recorded as changes of atmospheric pressure. See illustration on page 178.

This illustration also shows how the handle of the plunger is cut off about an inch above the rubber and sawed or filed flat on one side and a piece of fine emery cloth or

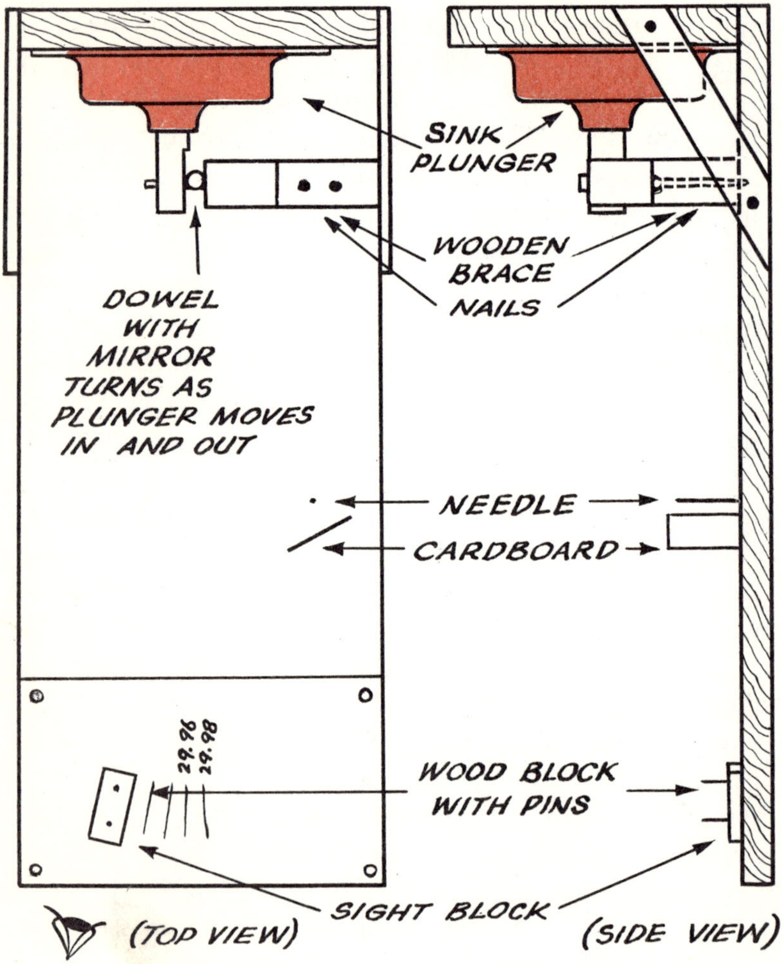
SINK PLUNGER
WOODEN BRACE
NAILS
DOWEL WITH MIRROR TURNS AS PLUNGER MOVES IN AND OUT
NEEDLE
CARDBOARD
29.96
29.98
WOOD BLOCK WITH PINS
SIGHT BLOCK
(TOP VIEW)
(SIDE VIEW)

sandpaper glued to the flat surface. Apply a liberal amount of glycerine to the bottom edge of the plunger and press it firmly onto the glass. Push it down as far as it will go. Then, with a knife, force up the bottom of the plunger to let in a small amount of air. The plunger handle will rise slightly. Quickly remove the knife. This provides enough room for the plunger to move in as well as out.

With the plunger in place you can get the measurements of the wooden brace. This can be made out of three-quarter-inch stock. It should be very sturdy. Glue an identical piece of emery cloth or sandpaper to the end surface. Do not fix the wooden brace to the base as yet. See illustration on page 178.

This illustration shows you how to prepare the mirror. Be careful when cutting the mirror, because, although getting small pieces of glass cut square and to the proper size is not difficult, you can cut your fingers in the process. The rod can be a short piece of quarter-inch dowel or a short piece from a toy construction set. One end is shaped as shown and the mirror connected to it after it has been scratched. The scratch on the back surface of the mirror is important. It should be deep enough to go clear through to the glass itself. This will make it easily visible from the mirrored side. The scratch can be made with a straight edge and a nail or other sharp point. When gluing the mirror onto the rod, make sure the scratch is centered on the rod and parallel with it. The mirror is then placed in position as shown on page 178.

Hold the rod, with the mirror on its lower end, in place between the rubber sink plunger and the wooden brace. Move the brace back and forth until you feel that the pressure of the plunger is enough to make a good contact with the rod. Mark the position of the brace and then nail it in place. Put the rod and mirror into position between the brace and the handle of the plunger. Gently push the end of the plunger inward and

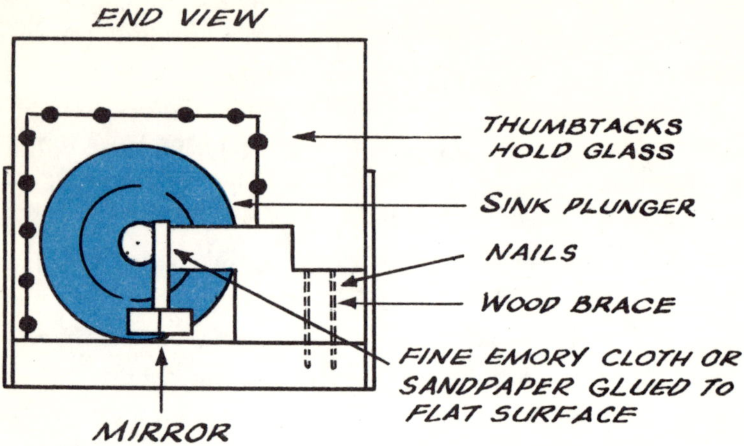
END VIEW
THUMBTACKS HOLD GLASS
SINK PLUNGER
NAILS
WOOD BRACE
FINE EMORY CLOTH OR SANDPAPER GLUED TO FLAT SURFACE
MIRROR

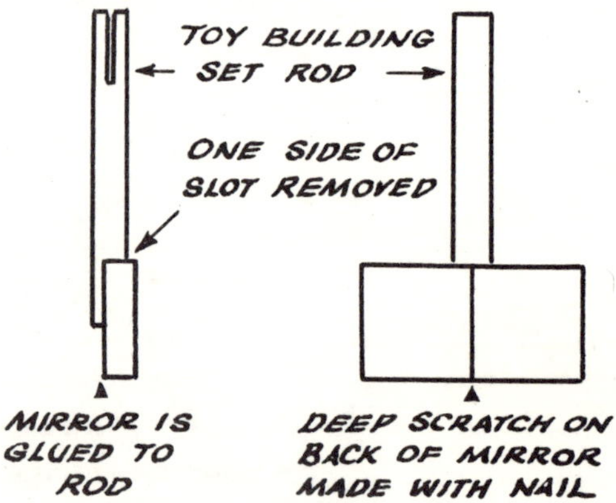
TOY BUILDING SET ROD
ONE SIDE OF SLOT REMOVED
MIRROR IS GLUED TO ROD
DEEP SCRATCH ON BACK OF MIRROR MADE WITH NAIL

1"
FILE FLAT
FINE EMORY CLOTH OR SANDPAPER GLUED TO FLAT SURFACE

watch the action of the mirror. It should rotate slightly and come back to its original position when you remove your finger.

Put a small needle into the baseboard as shown in the first figure in this chapter. Put a piece of white paper or cardboard on a block of wood behind the needle to make it more visible. Make a "sight block" out of half-inch stock with at least one edge straight, as shown here:

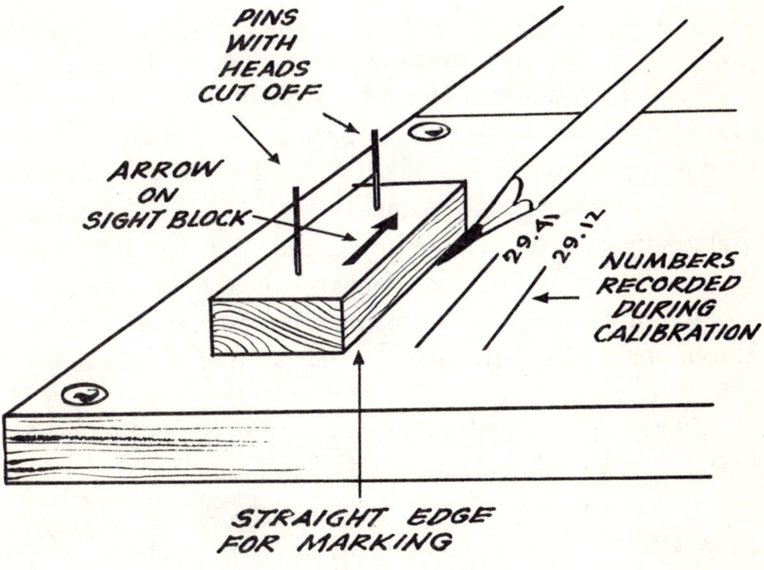

Put two pins in it as indicated and cut their heads off with a wire cutter. Draw an arrow on the top of the block and always use the block with the arrow pointing toward the mirror.

You are now ready to calibrate your barometer. To do this, sight down the center of the board into the mirror and turn the mirror until you can see the image of the needle. Make fine adjustments by turning the mirror until you get the image of the needle to fall exactly on the scratch on the mirror. Line up the two pins in the sight block with the scratch and the image of the needle. When they are all exactly in line, mark the right-hand edge of the sight block with a pencil.

Call up the Weather Bureau or consult a mercury barometer or a good aneroid barometer to find out what the correct barometric pressure is. It is usually stated in inches of mercury and will be a figure like 29.92. This is normal atmospheric pressure at sea level. Note the barometric pressure on the piece of paper tacked to the baseboard. Make readings every day at about the same time and record the results.

While the plunger is affected by changes in pressure, it is also affected by changes in temperature. If the air inside the plunger gets warm, it will expand. This will make your barometric readings lower than they should be. The readings will be higher if the temperature of the air inside the plunger drops. The rubber of the plunger will also be affected by temperature changes. For this reason, you should keep your barometer somewhere where the temperature does not change very much.

An interesting additional project might be to calibrate your barometer for differences in temperature.

To make your barometer work properly, you should understand why the mirror magnifies the movement of the plunger handle. See illustration on facing page.

Physicists have discovered that the angle at which the light from the needle falls on the mirror will be equal to the angle at which the light is reflected from the mirror. These two angles are indicated in the figure by A and B. When you line up the sight block with the scratch on the mirror and the image of the needle, you are in effect establishing the imaginary line that defines angle A. Note that the angles are

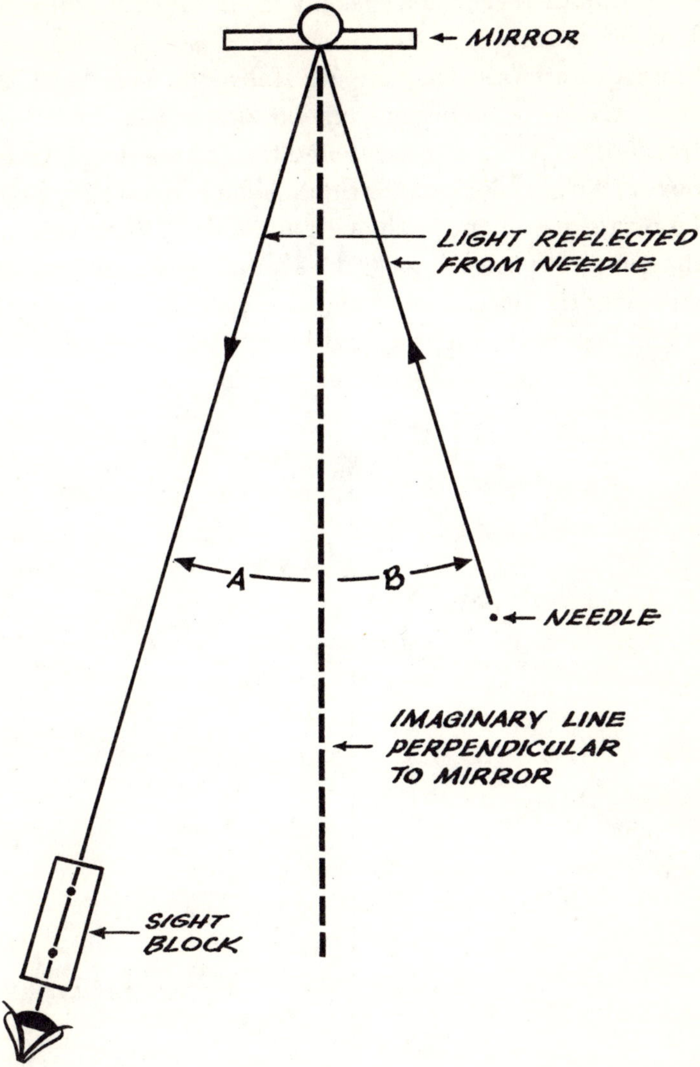
MIRROR
LIGHT REFLECTED FROM NEEDLE
A
B
NEEDLE
IMAGINARY LINE PERPENDICULAR TO MIRROR
SIGHT BLOCK

measured from the perpendicular to the mirror at the point where the light is reflected. This point is the scratch on the back of the mirror.

When the atmospheric pressure falls, the handle of the plunger moves out. This movement rotates the dowel and mirror slightly. Now, in order to find the new angle, you have to move the sight block to the right. The change in the plane of the mirror has changed the perpendicular. When you line up the pins on the sight block, the scratch on the mirror, and the image of the needle, you have established the new angle. This is indicated by angles C and D.

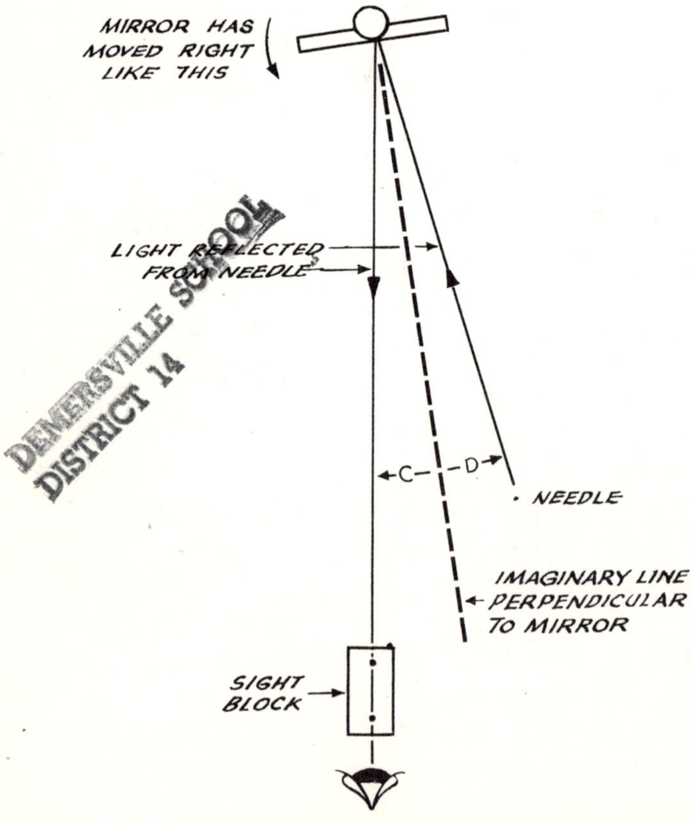

When the pressure gets greater, the mirror will rotate in the opposite direction and the angle you will be sighting will be greater.

Another way of thinking about the mirror and how it magnifies the movement is to think of it as a lever. If the plunger handle moves three quarters of an inch, a point on the circumference of the dowel will move this far also. With a quarter-inch diameter dowel, this will amount to slightly more than one full rotation! A tiny change in the angle of the mirror will be magnified by the distance at which you read the angle. The farther away you make the readings, the greater will be the distance included in the angle. In this way you can consider the mirror and sight lines to be an "optical lever."

That's how your barometer works. In order to use it to help you forecast the weather, you'll have to know why it behaves the way it does.

When air is heated, it expands . . . just as most other things do. This makes it weigh less than an equal volume of cooler air. Cool air, being heavier, pushes down harder against the surface of the earth. When the pressure is high it means that, more than likely, the air above is cool. Meteorologists also know that there is invisible moisture in the air. Cool air can hold less moisture than warm air. Therefore, when the barometer tells you that the pressure of the air is increasing, you can be pretty sure that cool, relatively dry air is over you.

If the air is cool and dry, it doesn't have much of the necessary moisture for rain or snow. This is one of the reasons a rising barometer indicates that cooler and fair weather is ahead.

When the barometer indicates that the pressure of the air is low, just the opposite can be expected. Low pressure usually means warm air is overhead, and warm air can hold more moisture. If this warm, moist air is cooled, the moisture will be released in the form of rain, snow, or sleet, depending upon the temperature of the air and other factors.

Now you're ready to try your hand at forecasting what the weather is going to be.

One of the safest predictions you can make is that bad weather is coming soon when your barometric pressure readings fall rapidly.

When you combine pressure changes with observations about the direction of the wind, you can get further clues to the coming weather. When the wind comes from the south or southeast and the barometer readings fall rapidly, a storm is probably coming from the west or northwest, and the center will pass near by within twelve to fourteen hours. When the wind comes from the east and northeast and the barometer readings fall steadily, a storm is probably coming from the south or southwest, and its center will pass near by within twelve to twenty-four hours. The wind in the latter case will shift to the northwest. The faster the barometer readings fall and the lower the reading, the more severe the storm will be.

As a good scientist, you know that you should keep accurate records. As a meteorologist, you should record the date, time, atmospheric pressure, air temperature, direction and force of the wind, and the clouds in the sky. The meteorologist at the United States Weather Bureau records all this information and much more. Before he makes a forecast, he studies these data along with reports from other weather stations. After you've had more experience, you can consult the published weather charts and see why he predicts the weather he does.

Meteorologists send weather information to the United States Weather Bureau from all over the world. Men on board ships and planes at sea send in readings. Technicians fly high in planes and send up special balloons carrying instruments to measure conditions in the upper atmosphere and radio back

Lightning over the George Washington Bridge was photographed as part of a study of the effect of lightning on high-tension wires. Meteorologists work with physicists and engineers on such a project. (Photograph courtesy of General Electric.)

information. In remote areas, scientists of the Weather Bureau have installed automatic weather stations that can radio back observations without anyone going near the stations for months.

Meteorologists at the main office of the Weather Bureau in Washington study these weather data and from it compile weather maps. Weather maps are also made in other major cities. The information on the weather map is in symbol form, and from it a trained meteorologist can find out quickly what conditions are in other areas and use this information to help him with his forecast.

These weather forecasts are much more important than you might think at first. You may use them only to help you decide what to wear to school or whether you can plan a picnic for the weekend. But businessmen rely on weather reports for planning their schedules. Is there a chance the open-air concert will be rained out? How about the baseball games? Snow removal means extra men must be hired.

Here are some of the activities that can be affected by the weather: constructing a house, tennis, hunting, road building, golf, state or county fairs, retail sales, boating, highway traffic, consumption of electric current for air-conditioners and refrigerators, purchase of seasonal clothing, etc.

Besides making observations and forecasts, meteorologists often study records to find ways of making more accurate forecasts and making them for longer periods ahead. And meteorologists are even experimenting with methods of controlling weather . . . at least part of it. You've probably heard about attempts to make rain. This was done first in 1946 by Vincent Schaefer, an American scientist. He dropped solid carbon dioxide particles into super-cooled clouds. This helped condensation take place and rain resulted. Schaefer and another American scientist, the late Irving Langmuir, made "artificial" snowstorms. Meteorologists have also tried other methods of "seeding" clouds to make rain.

As a beginning meteorologist, you won't be able to forecast the weather as well as the United States Weather Bureau scientists, but you can have a lot of fun finding out about "things in the air."